Information Systems Engineering and Management

Volume 67

The book series "Information Systems Engineering and Management" (ISEM) publishes innovative and original works in the various areas of planning, development, implementation, and management of information systems and technologies by enterprises, citizens, and society for the improvement of the socio-economic environment.

The series is multidisciplinary, focusing on technological, organizational, and social domains of information systems engineering and management. Manuscripts published in this book series focus on relevant problems and research in the planning, analysis, design, implementation, exploration, and management of all types of information systems and technologies. The series contains monographs, lecture notes, edited volumes, pedagogical and technical books as well as proceedings volumes.

Some topics/keywords to be considered in the ISEM book series are, but not limited to: Information Systems Planning; Information Systems Development; Exploration of Information Systems; Management of Information Systems; Blockchain Technology; Cloud Computing; Artificial Intelligence (AI) and Machine Learning; Big Data Analytics; Multimedia Systems; Computer Networks, Mobility and Pervasive Systems; IT Security, Ethics and Privacy; Cybersecurity; Digital Platforms and Services; Requirements Engineering; Software Engineering; Process and Knowledge Engineering; Security and Privacy Engineering, Autonomous Robotics; Human-Computer Interaction; Marketing and Information; Tourism and Information; Finance and Value; Decisions and Risk; Innovation and Projects; Strategy and People.

Indexed by Google Scholar. All books published in the series are submitted for consideration in the Web of Science.

For book or proceedings proposals please contact Alvaro Rocha (amrrocha@gmail.com).

XinYing Chew · Abbas Gatea Atiyah ·
Alhamzah Alnoor · Sammar Abbas ·
Yousif Raad Muhsen · Gül Erkol Bayram

Partial Least Squares Structural Equation Modeling and Complementary Methods in Business Research

Springer

XinYing Chew
School of Computer Sciences
University Sains Malaysia
Pinang, Malaysia

Alhamzah Alnoor
Department of Management
College of Business Administration
A'Sharqiyah University
Ibra, Oman

School of Computer Sciences
University Sains Malaysia
Pinang, Malaysia

Yousif Raad Muhsen
Department of Computer Science
College of Computer Science
and Information Technology
Wasit University
Wasit, Iraq

Abbas Gatea Atiyah
Faculty of Administration and Economics
University of Thi-Qar
Nasiriyah, Iraq

Sammar Abbas
College of Business and Entrepreneurship
Abdullah Al Salem University
Kuwait City, Kuwait

Institute of Business Studies
Kohat University of Science
and Technology
Kohat, Pakistan

Gül Erkol Bayram
School of Tourism and Hospitality
Management
Sinop University
Sinop, Türkiye

ISSN 3004-958X ISSN 3004-9598 (electronic)
Information Systems Engineering and Management
ISBN 978-3-032-01054-4 ISBN 978-3-032-01055-1 (eBook)
https://doi.org/10.1007/978-3-032-01055-1

This Springer imprint is published by the registered company Springer Nature Switzerland AG
The registered company address is: Gewerbestrasse 11, 6330 Cham, Switzerland

Preface

Historically, business and other social sciences research has been dominated by quantitative research, where researchers deduce hypotheses from literature, develop statistical models, and test them using a variety of quantitative or statistical techniques. Partial Least Squares Structural Equation Modeling (PLS-SEM) has emerged as a prominent statistical method for analyzing the complex relationships between variables. Over the past few years, applications of PLS-SEM have grown unprecedentedly with the inclusion of some novelties such as the use of Heterotrait-monotrait criteria to assess the discriminant validity of the construct, methods to measure unobserved heterogeneity, and assessment of measurement model invariance. PLS-SEM is particularly useful in instances of predictive modeling, complex models, limited sample size, and non-normally distributed data. In the context of business research, researchers need to explore the complex relationships between multiple variables in a single model. Similarly, obtaining a large sample in business studies is quite a difficult task because of the access issues and other ethical concerns. PLS-SEM is less sensitive to sample size and works well with small sample sizes. Business research is more inclined to forecasting various business activities and making decisions accordingly. Predictive power of PLS-SEM makes it a more appropriate tool for business researchers, aligning with the needs of applied business research. In the majority of cases, the constructs used in business research are formative ones. PLS-SEM is perfectly suitable for formative measurement models. Contemporary business research is more concerned with evolving issues, such as digitalization, new technology adoption, the metaverse, social commerce, sustainability, etc. Literature is not much developed in these domains, and researchers have to rely on limited available literature. PLS-SEM is useful for such type of exploratory research and helps to identify the key indicators/variables and to determine the underlying relationships. Apart from this, PLS-SEM facilitates group comparison (male vs. female employees, urban vs. rural population, national vs. international markets, local vs. foreign companies, etc.). Added benefits of mediation and moderation analysis make PLS-SEM a preferred choice for business research.

Though there are numerous benefits of using PLS-SEM in business research but at the same time, it encounters some major potential drawbacks. Researchers cannot fully rely on PLS-SEM as a sole choice to obtain rigor in the findings. For instance, PLS-SEM is less suitable when the assumed relationships are derived from well-established theory(ies). Business research is not always interested in prediction; rather, confirmation is required, coupled with model fitness and rigor. PLS-SEM can only establish the relationship between the observed variables and constructs, but it lacks the power to establish the priorities and strength of influence of observed variables separately. Sometimes, researchers are also interested in providing a solution to the problem by establishing the necessary conditions for the solution. PLS-SEM is not a valid technique in this case. PLS-SEM has a weakness in handling large datasets, and researchers need to rely on other techniques, such as covariance-based structural equation modeling (CB-SEM), also suitable for confirmatory modeling. Another drawback of PLS-SEM is that it relies on bootstrapping for establishing the significance of the relationship. Bootstrapping is quite complex, with some important assumptions to meet. Researchers occasionally overlook these assumptions, causing the instability of findings. PLS-SEM is not recommended when the latent variables are highly correlated because it cannot distinguish between these constructs. If used, it can cause a multicollinearity problem and unclear paths. PLS-SEM has its limitations, and the avoidance of those limitations leads to false and ambiguous results.

The authors are convinced that PLS-SEM cannot fully serve the purpose of meaningful research on its own unless it is combined with other techniques for better and more reliable results. Accordingly, authors recommend the use of other complementary techniques that overcome the above-mentioned potential drawbacks of PLS-SEM. Authors are highly convinced that use of other complementary techniques such as artificial neural networks (ANN), fuzzy-set qualitative comparative analysis (fsQCA), multi-criteria decision making (MCDM), and adaptive neuro fuzzy inference system (ANFIS) can fully comprehend the overarching research objectives and make the research a worthwhile effort. For instance, ANN can make more accurate predictions and inferences involving complex relationships through iteration. FsQCA uses fuzzy logic to suggest a better combination of conditions (variables to exist) for the proposed outcome. MCDM is used to choose the best alternative among the competing alternatives. Similarly, ANIFS combines ANN and FsQCA for better results.

The significance and importance of PLS-SEM as an evolving statistical modeling technique cannot be denied, and there is plenty of research that has relied on it. However, with the growing interest in the technique, it has become obligatory to introduce the technique with all its complexities in a simple, understandable, and linear way, particularly for the novice and early career researchers. The authors of this book have felt this need after years of experience in teaching and doing research. The purpose of this book is to dispense the understanding and application of the above-mentioned techniques in a simple way in the field of business studies. The authors tried to capture the issues encountered by the researchers using PLS-SEM and other complementary techniques. We expect that this book will initiate a new debate about the extended applications of PLS-SEM as a fundamental tool for business research. Considering the emerging nature of business phenomena involving complex latent constructs, we also believe that this book will be a useful addition to the already existing stock of knowledge in the domain. We have tried to cover a wide range of topics to make our work more useful for diverse and wider readers. However, we also understand that we may have omitted some important considerations, for which we will encourage and appreciate critical commentary from our readers.

Pinang, Malaysia — XinYing Chew
Nasiriyah, Iraq — Abbas Gatea Atiyah
Ibra, Oman/Pinang, Malaysia — Alhamzah Alnoor
Kuwait City, Kuwait/Kohat, Pakistan — Sammar Abbas
Wasit, Iraq — Yousif Raad Muhsen
Sinop, Türkiye — Gül Erkol Bayram

Competing Interests The authors have no competing interests to declare that are relevant to the content of this manuscript.

Contents

About the Authors

Dr. XinYing Chew is Associate Professor in the School of Computer Sciences, Universiti Sains Malaysia. She is also Program Manager of the School of Computer Sciences, Universiti Sains Malaysia. She holds a Ph.D. in statistical quality control from University Sains Malaysia. Her area of interest is in advanced analytics and statistical quality/process control. She is a certified trainer with Human Resources Development Fund (HRDF). She is also the trainer for MDEC-Intel AI Academy Program and SAP Next-Gen Program. She is Adjunct Research Fellow of Swinburne University of Technology, Sarawak Campus. She is also Professional Technologist with the Malaysia Board of Technologist (MBOT). Prior to her academic career, she worked in the advanced analytics research team of a renown US multinational company. Her areas of research are in advanced analytics and statistical quality/process control.

Dr. Abbas Gatea Atiyah is Associate Professor of organization studies at the Faculty of Management and Economics, University of Thi-Qar, Thi-Qar, Iraq. He received his Ph.D. in business administration from University of Karbala, Iraq. His research interests center on organizational studies and strategic management. He published many papers and book chapters in different journals and proceedings. He is a reviewer of many journals.

Dr. Alhamzah Alnoor is Assistant Professor at A'Sharqiyah University (ASU) in Oman, specializing in Strategic Management and Operations Research. He earned his M.Sc. with First Class Honors from Southern Technical University in 2013, followed by an MBA from the University of Basra in 2016. Dr. Alnoor completed his Ph.D. at the Faculty of Management, University of Science Malaysia, in 2022. Additionally, he received a Postdoctoral Fellowship from the School of Computer Science at Universiti Sains Malaysia in Penang in 2024. His research interests encompass strategic management, business ethics, leadership, human resource management, organizational psychology, organizational sociology, innovation, entrepreneurship, small and medium-sized enterprises (SMEs), technology adaptation, and social commerce. Dr. Alnoor has published over 100 research articles in journals indexed by WOS and Scopus, with publications in high-quality journals such as *Technological Forecasting and Social Change*, *Journal of Retailing and Consumer Services*, and *Business Strategy and the Environment*. He also serves as an editor for various journals under Wiley, ScienceDirect, and Springer. Additionally, he is a member of the editorial boards of several prestigious academic journals and has authored multiple books published by Springer and Emerald. Notably, Dr. Alnoor was ranked among the top 2% of scientists based on Stanford University's 2024 metrics.

Dr. Sammar Abbas is Associate Professor of organization studies at the Kohat University of Science and Technology in Kohat, Pakistan, and College of Business and Entrepreneurship, Abdullah Al Salem University Kuwait. He received his doctoral degree in business administration at the University of Essex in the UK. His research interests center on organizational studies. He has published many papers in high-quality journals.

Dr. Yousif Raad Muhsen is Associate Professor at College of Engineering, Wasit University, Wasit, Iraq. He is a researcher, who has published research in various journals. He serves as a reviewer for several journals. He received his Ph.D. from Department of Computer Science, Faculty of Computer Science and Information Technology, Universiti Putra Malaysia, Seri Kembangan, Malaysia.

Dr. Gül Erkol Bayram is currently Associate Professor in the School of Tourism and Hotel Management, Department of Tour Guiding, University of Sinop, Sinop, Türkiye. Dr. Erkol Bayram has worked as an internal trainer and teacher in the tour guiding arena. Her doctorate is in Tourism Management from the Sakarya University, Türkiye, and she completed her dissertation research on Tour Guiding in Türkiye. Her core subjects are tourism, tour guiding, tourism policy, and planning, women studies. Erkol Bayram has also worked as a professional tour guide in the tourism sector. The editor has many book chapters in the international arena and published her books as an editor in the national arena. She has edited a book edited on *Tour Guiding-Past, Present, Future* under Detay Publishing, Türkiye, *Women and Tourism* under Detay Publishing, Türkiye. Also, she has published a book named *Dynamics of Tourism Industry: Post Pandemics and Disasters* under Apple Academic Press (CRC Press a Taylor and Francis Group). Currently, she is working on edited book named *Resilient and Sustainable Destinations After Disaster: Challenges and Strategies* under Emerald Publishing. She also has many book chapters under Emerald, CABI, IGI Global also. She has been invited for many talks/lectures/panel discussions by different Universities like University of Mumbai, India, International Multidisciplinary Conference, College for Women, Parade Ground, Jammu, India, International Faculty Development Programme, International Digital Conference 2021 on Curriculum 4.0., Jagdish Sheth School of Management and University of Management and Technology, Lahore, Alhamd Conference on Education, Islamic University, Islamabad, Pakistan, etc. The government of India has invited her to deliver a talk on Tour Guiding and the University of Kota about a career in tourism, Taylor's

University on Tour Guiding. Dr. Erkol Bayram was a special guest, guest of honour to many conferences, organizations. She got awarded from several organizations. Erkol Bayram has also worked as a professional tour guide in Türkiye.

Part I
Introduction

Chapter 1
Partial Least Squares Structural Equation Modeling

Learning Objectives

1. Understand the foundations and applications of PLS-SEM.
2. Differentiate between reflective and formative measurement models.
3. Identify and address key methodological issues in PLS-SEM.

1.1 Introduction: PLS-SEM

Partial Least Squares (PLS), also known as PLS-SEM (Structural Equation Modeling), was introduced by Wold [1] with the intended use in econometrics and to address the weak theory and inconsistent, incorrect data issues. Recently, it has become an established statistical method widely used across social sciences research to determine the relationship between multiple variables. Social science research has predominantly remained under the quantitative tradition, with a focus on defining the variables and their constructs and then testing the inter-relationships between them [2, 3]. It is useful in instances when the sample size is relatively small and there are some missing values or a multicollinearity problem. Thus, it overcomes the issues of missing values, multicollinearity, and unstable results obtained using the ordinary least squares method (OLS) when used for small sample sizes [4, 5]. The name 'partial' has been derived due to the characteristic feature of the technique that it 'partially' extracts the data for a small sample, opposite to covariance-based SEM (CB-SEM, which requires a relatively large sample size to make inferences. The common purpose is to determine the outcome variable (Y) from a predictor (X) and to describe the underlying relationship between the two variables [5, 6]. In the majority of cases, the predictor is a latent variable composed of other directly measurable variables that are related to the outcome or response variables. SEM is fit for capturing the relationship between latent variables and is also called latent variable analysis [7].

X. Chew et al., *Partial Least Squares Structural Equation Modeling and Complementary Methods in Business Research*, Information Systems Engineering and Management 67, https://doi.org/10.1007/978-3-032-01055-1_1

PLS-SEM is being increasingly applied as more researchers are becoming aware of the usefulness of the technique for determining and evaluating complex relationships [8]. Magno et al. [9] conducted a comprehensive review of using PSL-SEM and found that the main reasons for using the technique include small sample size and normal data. With a small sample size, PLS-SEM applies multiple forward–backward iterations to achieve optimum predictions [8]. However, Hair et al. [8] argued that the choice of using the technique must be based on the purpose of the study. Becker et al. [10] suggested that the purpose of choosing a technique should not only be determining the relationship between the variables, but it should also be capable of making predictions and recommendations for a practical purpose. PLS-SEM serves both purposes. PLS-SEM fits with the recent research trends because it offers a balance between the explanation of underlying relationships and the predictions for practitioners. The most common reasons for using PLS-SEM include, but are not limited to, model complexity, theory development and exploration, and making predictions [8, 9]. Regarding model complexity, PLS-SEM can better capture complex non-linear relationships involving more predictors, constructs, response variables, moderators, and mediators. Hair and Alamer [8] suggested that the decision to use PLS-SEM is dependent on multiple factors, including the objectives of the study (explanation & prediction), measurement model type (composite constructs), sample size (smaller one), data distribution (no normal distribution), model complexity (multiple variables), and higher order construct with more refinement of the constructs. Ringle et al. [11] argued that a clear explanation must be provided regarding the use of PSL-SEM in meeting the study objectives. Dash and Paul [7] described that SEM better captures dependence and interdependence relationships among the constructs of the main variables and provides variance-based analysis. It also helps to choose the model fit among the alternative models that is appropriate to capture the suggested relationships.

This chapter introduces different aspects of PLS-SEM and provides a basic understanding of the technique for novice researchers. The chapter is divided into seven sections. Section two explains different measurement models, and section three discusses reflective vs formative models. In section four, the issue of common bias is discussed, whereas section five discusses common method variance. Section six explains marker variables, and the seventh section is about the lateral collinearity.

1.2 Measurement Models: Reflective Versus Formative

A measurement model is a general model that reflects the relationship between latent variables and its items or indicators. Literature on reflective and formative measurement models mentions two perspectives about the two models [12]. The first perspective is closer to realist ontology that a 'construct exists' independent of our assessment of the construct, and the second perspective suggests that constructs do not exist, rather they are formulated through indicators i.e., they require assessment [13]. Chang et al. [12] referred to them as reflective and formative constructs or models, respectively. Reflective generally means that a change in one variable reflects the change in another variable [14]. In social science and management research, researchers determine the statistical relationship between the latent variable and its indicators by measuring correlation or covariation. For instance, if the variation in the indicator is linked with the variation in the latent variable, then any external factor that changes the latent variable is supposed to necessarily influence the indicators [12]. In simple words, any exogenous intervention disrupting the latent variable must be reflected in the indicators. The relationship between the latent variable and its indicators is reflective. Any change in the latent necessarily brings change in indicators, thus a kind of causal (cause-effect) relationship between the two, where the causality flows from the latent to the indicators [15]. Some scholars argue that not all the latent variables need to be formed with the combination of positively related indicators [16]. Alternatively, latent variables can be formed with the combination of indicators without assuming the defined relational pattern between them. This means that any change in the indicators will be reflected in the latent. There is also a causal relationship between the two, but the direction of causality is from the indicators to the latent [17]. It is critically important to differentiate between reflective and formative measures for model specification and deriving meaningful relationships between the variables. Coltman et al. [15] (p. 1252) presented a comprehensive framework for the assessment of reflective and formative models. The framework is based on the earlier studies that differentiate between the two models across different dimensions. These dimensions include the nature of latent variables [s], direction of causal relation, characteristics of items used to determine a construct, correlation between items, relationship between the antecedents and consequences of a construct, and items.

Regarding the nature of the latent variable, in the case of the reflective model, the latent variable exists independently of its indicators, whereas for the formative model, the latent variable is formed from its indicators [15, 18]. As far as the causal relationship is concerned, in the case of the reflective model, the change in the latent variable causes change in the indicators, whereas in the case of the formative model, change in the latent variable does not necessarily cause change in the indicators. But if there is a change in the value of indicators, that does not influence the value of latent variable if there is case of reflective model, but for formative model, a change in the value of indicators leads to change in the value of latent variable [19, 20]. Concerning the item's characteristics, if the items can be replaced with other items but this does not influence the theoretical underpinning of the concept (latent variable) when there is reflective model, but in case of formative model, the case is different where the items cannot be replaced and adding or dropping items disturbs the theoretical understanding of the concept. Items in reflective models are positively correlated, whereas in the case of the formative model, a negative relationship between the items is also possible [12, 15]. For the relationship between the items and the latent variable, all the items have a similar level of relational significance, whereas the level of relational significance between the items and the latent variable may be different from one item to another and the latent variable.

1.2.1 Assessment of Reflective and Formative Models

To make the assessment of the two models, researchers recommend different criteria. In the case of reflective model assessment, internal consistency reliability, indicator reliability, convergent validity, and discriminant validity are used [21]. Internal consistency reliability indicates that the items are related to the latent variable. The usual measure is Cronbach's alpha with an acceptable value ≥0.7 [22]. Indicator reliability indicates the suitability of indicators to define the latent variable or the construct and is determined through indicator loading with an acceptable value of >0.708 [23]. Convergent validity means whether the different indicators used to measure a construct provide the same measure, i.e. statistical verification of theoretically related indicators to the construct [24, 25]. The value of average variance extracted should be >0.5 to confirm the convergent validity among the different measures of a construct [21, 24]. Discriminant validity is a statistical measure that confirms that the constructs (usually unobservable), which are not conceptually related, are also not statistically related. Heterotrait-monotrait ratio of correlations (HTMT) is a common measure of discriminant validity with the threshold value of >0.85 [26] or some suggest >0.9 [27]. For the assessment of a formative model, three common measures, i.e., multicollinearity, construct validity, and indicator reliability, are used. Multicollinearity refers to the extent to which the indicators relevant to a construct are interrelated. A high degree of multicollinearity makes it difficult to isolate the effect of each indicator on the response variable [28]. Variance inflation factor (VIF) is a common measure to estimate multicollinearity, and the ideal value of VIF is <5. A VIF value = 1 indicates the absence of a multicollinearity problem [29]. Construct validity refers to the extent to which a construct measures what it intends to measure. Construct validity can be assessed through convergent validity and discriminant validity [21].

1.2.2 Common Method Bias

Social science quantitative studies rely on a self-administered questionnaire as a data collection tool. Researchers use an already established measurement scale or develop a new scale consisting of constituent items defining the variables. Occasionally, the instructions provided to fill the questionnaire may influence respondents' responses in the same general directions, thus causing the common variation across indicators or over-representation of certain factors [30]. Schwarz et al. [31] argue that CBM occurs when a common method is used to obtain the different measures of the same construct or trait. Though the use of a questionnaire is a common data collection technique, it can result in common method bias and thus cause doubts about the reliability and validity of the results [32, 33]. Common method bias occurs when the same scale measures the response against indicators and response variables. Also, the desirability to obtain specific responses causes common method bias. Another common factor that causes common method bias is the researcher himself/herself, especially when known to the respondents, the values and beliefs of the researcher can influence the responses [34]. Due to the prevalence of the issue, scholars have a wider debate on how to identify and avoid common method bias [12, 35, 36]. Hence, it is critically important to test the variables and constructs for common method bias. Social science scholars have moved beyond the debate about the occurrence of CBM, there is general agreement about its existence. Now, scholars are more interested in issues like identification, magnitude, and the avoidance of CBM [31]

To deal with the issue of CBM, the first step is to identify its sources. Schwarz et al. [31] mentioned six major sources that inject CMB and include ambiguity and complexity of the items, choice and format of the measurement scale, negatively worded questions, and inclusion of items to influence the response, items' alignment with respondents' traits, and influence of the external occurrences on respondents' responses [37]. Regarding the ambiguity and complexity of the items used, researchers may include questions that are not worded, have double meanings, or have different meanings in different settings [38]. Use of technical words or uncommon words can also lead to ambiguity and complexity on the part of respondents [31]. To answer such questions, respondents can draw unique meanings that lead to a lack of internal consistency among the responses and thus cause the problem of construct reliability. Similarly, the format of scale or scale anchoring also influences responses because respondents are more focused on scale anchors rather than the individual items. For instance, the use of scale anchors (strongly disagree—strongly agree) is facilitating for the respondents but can divert their attention from the question items to the scale anchors. This leads to CBM because it can generate patterned responses depending on the format of the questionnaire [31, 36]. In case of negatively worded questions, respondents may misinterpret them, and this creates weaker factor loading, i.e., reduced construct reliability. In the same way, the priming effects, i.e., explanation of the items and their intended purpose, can also influence the responses, resulting in items' face validity and construct reliability. Respondents' responses to different items are also dependent on the way they correlate items with their traits, causing a more positive or more negative response. Lastly, transient mood state also affects response in a particular way, i.e., after passing through a specific transient state, the respondents have a particular mood, and that significantly his/her responses to the items [31, 38].

Once the sources of CBM are identified, the next step is to provide remedial measures. Scholars argue that there are two broad remedies to treat CBM. The first one is carefully designing the measurement scale, and the second one is the use of statistical techniques [33, 39]. The first one is applied at pre pre-data collection stage, and the second one is applied at post post-data collection stage. It is recommended to use both to avoid the CBM issue and thus to enhance the reliability of the construct for robustness of the results [40]. Podsakoff et al. [41] and Tehseen et al. [42] listed down remedial measures in this regard. First, the researcher should carefully design the survey questionnaire by considering the above-mentioned sources, such as the use of double-barreled questions, ambiguous words, priming effects, etc. Researchers must obtain data from different sources because it will control respondents' bias towards the relationship between the predictors and outcomes. However, this can be a costly and time-consuming method and may also involve multiple subjective issues [41]. The other method is the separation of the responses collection on the independent and dependent variables, i.e., the data on independent and dependent variables can be collected at different times or making it explicit to the respondents that the variables are not related to each other. This will help to get the 'forced responses or biased ones. Separation of the items from the context can also help to avoid method bias [39, 41]. Careful construction of the scale with the involvement of the experts can enhance the reliability of the construct. These are not the only available procedural remedies to deal with CBM, researchers can look for other subjective measures depending on the time and context of data collection. However, it must be noted that all the above procedural remedies may not be workable simultaneously or not applicable in different situations and contexts, and scholars are in a better position to decide about the choice of these measures [42].

Beyond these subjective or procedural remedies, there are also statistical techniques that can be applied to deal with the CBM problem. It is very critical to measure responses against each item correctly to enhance the theoretical and managerial implications of a research study. Also, if the CBM is not addressed, the generalizability of the findings becomes doubtful and compromised [43]. Additionally, the existence of CBM results in the inaccurate measurement of the impact and relationship between the variables, resulting in serious consequences in terms of acceptance of the results. Therefore, it is recommended to ensure that both procedural and statistical remedies are well placed to mitigate the occurrence of CBM. Different statistical techniques are available to treat the CBM issue, but again, it is dependent on the researcher to make an appropriate choice. For instance, Liang et al. [44] developed 'Common Latent Factor (CLF) to address CBM. CLF considers all the indicators and their variance regarding a latent variable and thus avoids CBM [38]. The most common and widely used statistical techniques to outroot CBM include Harman's Single Factor Test, Partial Correlation Procedures, Correlation Matrix Procedure, and the Measured Latent Marker Variable Approach (i.e., construct-level and item-level approach). In the following, we describe the use of the above-mentioned statistical techniques.

1.2.3 Harman's Single Factor Test

Harman's Single Factor Test is a commonly used method to identify whether a single factor is responsible for the variance in the data. Using this method involves loading all the items for each construct into factor analysis to see if a single factor causes variance among the measures. If there is no single factor causing the variance, this indicates the absence of CBM [45]. To carry out the test, the following steps are followed [42].

1. Entering all the construct-related items into factor analysis and conducting Principal Component Analysis (PCA). This is done to convert larger data into smaller

data by reducing the dimensions of each construct (see Abdi and Willaiams [46] for the details).
2. Count on the factor unrotated solution to know the factor causing variance; it locates the pattern of relationship in the data. Eigenvalues are calculated to extract the factors causing variance. Eigenvalues reflect the amount of variance caused by a factor.
3. See if the single factor emerges, causing the variance

While using the above test, two assumptions are made to ensure the absence of CBM. First, no single unrotated factor emerges that causes the majority of variance, and the first factor does not account for a higher percentage of the variance [40]. Though this test is commonly used for the identification of variance, it is not useful in controlling or correcting CBM [42]. The other statistical technique is the Partial Correlation Procedure (PCP).

1.3 Partial Correlation Procedures

Another statistical technique to control CBM is the Partial Correlation Procedures, which includes two statistical procedures, i.e., Partialling Out of General Factor and Partialling Out of Marker Variable (this will be discussed later in this chapter). Broadly, both techniques are similar, but there is an inherent difference between the two. The first one relies on the value of the first unrotated factor, whereas the second one incorporates the concept of social desirability. Social desirability is the demonstration of socially desirable behavior. The social desirability concept suggests that while responding to the questionnaire items, respondents choose the answer that is socially desirable, irrespective of the true representation of their true response. Social desirability is one of the major causes of CBM and causes the issue of construct reliability [47, 48]. Many of the researchers use Crowne and Marlowe's [49] 3-item social desirability scale to identify the variance caused by the social desirability factor. There are other scales consisting of fewer items and are equally useful, such as those developed by Fischer and Fick [47], Reynolds [50], and Ramanaiah et al. [51]. To perform Partialling Out of General Factor, Ramanaiah et al. [42] described the following 5-step procedure.

1. Calculate factor scores across all responses on 1st unrotated factor and create a separate data file to transfer the scores.
2. Add the scores calculated in 1st step (create another variable called general factor) to the cases file for the data set and transfer this file to Smart PSL software.
3. Construct a research model and observe the R^2 value for the endogenous variable.
4. Add a general factor to the endogenous variable, and again note the value of R^2
5. Calculate the difference between the observed value of R^2 calculated in steps 3 and 4.
6. A significant positive difference indicates the existence of CMB and vice versa.

To apply Partialling Out of Marker Variable, the following 3-step procedure is suggested.

1. Construct a study model with software (i.e., Smart PSL) and partial out the Marker Variable, i.e., social desirability.
2. Determine the correlation between the marker variable and the latent variable using the PLS algorithm for assessing CMB.
3. If the value of correlation calculated in step 2 is more than 0.3, this indicates the existence of CMB.

Podsakoff et al. [41] developed another method based on the difference between the R^2 values before and after introducing the Marker variable in the endogenous variable. A significant difference in the R^2 value for any of the endogenous variables denoted a higher CMB. For further details on the use of partialling general factor and the marker variable, see Tehseen et al. [42].

1.3.1 Correlation Matrix Procedure

A CMB issue exists when there is a high correlation among the main constructs of the study, i.e., the value of the correlation coefficient (r) is more than 0.90 [52]. To detect the value of 'r', at first, study model is constructed with Smart PLS software. After this, the correlation between the latent variables is observed using the PLS algorithm. The above-mentioned criterion for the 'r' value is applied to confirm the existence or non-existence of CMB. Same as Harman's Single-Factor Test, this technique also only identifies the presence of CMB and does not offer any corrective action. Another statistical technique, i.e., the Measured Latent Marker Variable (MLMV), is more suitable for dealing with Common Method Variance (CMV) and will be discussed in the relevant section.

1.3.2 Lateral Collinearity

Collinearity is a multivariate notion, where two or more variables measure the same attribute of an object or construct [53]. For example, consider two variables, "the net income of a person" and "social support payment" received by that person; the two variables may explain the spending pattern of that person. The concept of collinearity is often confused with correlation. Two variables are said to be correlated if they vary in concert with each other, even though the variables may measure different object attributes, e.g. a person's weight from 1 to 20 years of age and the price of gasoline over those years [53]. The difference between correlation and collinearity is that in collinearity, the variables are seen as referring to the same object attribute, i.e., in our example, total disposable income available to a person for consumption. Whereas in correlation, the variables have apparently no similar object attribute.

Correlation is useful only when there are two variables, also known as a bivariate phenomenon, whereas collinearity can be used for more than two variables, also known as a multivariate phenomenon.

"Lateral Collinearity is predictor-criterion collinearity, and it occurs when two variables that are hypothesized to be causally related measure the same construct. In a causal model, the two variables would be shown as linked by an arrow with one pointing to the other. In contrast to the lateral collinearity, there is a vertical or classical collinearity, which is a possible predictor-predictor redundancy phenomenon." [53]. Unlike vertical collinearity, lateral collinearity is rarely explicitly tested in multivariate analysis. Lateral collinearity can lead to misleading results because it can be masked by the appearance of a strong causal effect in the model. E.g., [54] while studying the relationship between age, gender, and ethnicity on eGFR (estimated Glomerular Filtration Rate) used to predict renal function found that increasing recipient age and reducing eGFR was artificially created by the equation used to estimate GFR. This is an example of lateral collinearity, which should be suspected when we derive the outcome from the predictors. This relationship is not identified when performing traditional collinearity measures such as VIF (Variance Inflation Factor) [54].

1.3.3 Why is Lateral Collinearity a Problem?

Although it is desirable to have some degree of collinearity in the model but too much collinearity in the independent variables makes coefficient estimates unreliable. The situation is problematic because lateral collinearity may exist even when correlation among variables in multivariate analyses is relatively low [53]. Moreover, Lateral collinearity also has a negative hidden impact on theory development because it allows researchers to find strong but artificial support for hypothesized causal relationships [53, 55].

1.3.4 Signs of Collinearity

Signs of Collinearity include high R-squared Coefficients among latent variables, loadings greater than 1 after nonnormalized oblique rotation in a confirmatory factor analysis. However, high R-squared Coefficients may not necessarily be caused by lateral collinearity but rather by vertical collinearity. High Correlation among latent variables exists. Correlation is a necessary but insufficient condition for collinearity to exist [56].

1.3.5 How to Deal with Lateral Collinearity?

Kock and Lyn [53] described the following procedure to deal with the issue of lateral collinearity.

1. It is recommended to perform full collinearity tests as part of routine data validation tests in research employing the variance-based Structural Equation Model.
2. Removal of Indicator. A pair of latent variables may be associated with a few offending indicators that are highly correlated with both latent variable scores.
3. Re-assign the indicator. It may be the case that an indicator may be associated with a variable to which it is poorly correlated and highly correlated with another latent variable. So, such an indicator should be reassigned to the latent variable to which it is highly correlated.
4. Latent variable removal. It may be the case that collinearity may arise due to all of the indicators of one or the other, or both variables being highly correlated with both latent variable scores. If that's the case, the latent variable should be removed.
5. Variable Aggregation. Where collinearity between a pair of reflective latent variables is due to a high correlation between all of the indicators of one or the other or both variables being highly correlated with both latent variable scores. The solution is to aggregate all of the indicators of the collinear latent variables into one latent variable.
6. Lateral collinearity should be avoided, in the first place, by designing the measurement instrument correctly. If a questionnaire is poorly designed, the techniques discussed above may not salvage the resulting data set. However, a properly designed questionnaire is also not a guarantee that the collinearity will be avoided due to different understandings about the question statement of the respondent and the designer of the questionnaire.

1.4 Marker Variable

Marker variable is explained as a variable in a study that is theoretically not related to at least one of the variables that are substantive and specifically included to spot and control CMV. Marker variable was first proposed by Lindell and Whitney [57], which pay attention on questionnaire data sets that contain a marker variable, which they explain as a variable that theoretically distinct from substantive variable and for which its probable correlation with substantive variable is zero and which constructed to be utilized to detect CMV in their correlation marker technique and So far those who potently recommend this method alert that their efficacy is probably a function of the marker quality used to apply them. They argue that the MV method partial out the consequences of common method variance from the correlation attained in mono-method cross-sectional study patterns and returns correlation values that are not contaminated by common method variance. They propose numerous criteria for the selection of MV correlation. In data, it provides an effective and efficient mean of post hoc detection of common method variance (CMV). Marker variables are measured in the same technique as substantive variables, but since they are conceptually distinct to the variables of interest, they are thought to be the proxy for common method variance (CMV) Leonard at el. [58]. The technique of marker variable proposes an easy and practical method to tackle the validity threats on account of CMV to the findings of the individual studies. Main determinants of a quality of marker variables are the level to which it is affected by the same effects of common method variance (e.g. acquiescence, effectivity) as a set of variables that is substantive but then it is not theoretically related to those substantive variables. To identify this type of marker variable is not a simple attempt, it is very hard to understand whether some variable actually reflects variance as a result to method and, if yes, then what grounds that method variance. Even though markers should be selected on theoretical and methodological points, extant theory is frequently insufficient for creating when variables will be not related to one another or to which sort of CMV will be utmost weak [59]. Market variable example includes questions which are demographic, attitude items that are not related or methods of construct that are conceptually far from the variables which are primary. After selecting an appropriate marker variable different techniques can be used to detect and control common method variance. Basic method examines the correlation among the marker variable and substantive variable, and significant correlation propose the method bias presence [57]. Partial correlation technique can be used for better modifications, where the effect of marker is statistically partial out from the association between substantive variables. Maker variable can be included as latent method factor or exogenous variable in the context of SEM, which allows the immediate estimation of the substantive relationship and the method bias [39]. For more precise representation of relationship between constructs structural model can be modified to account for the variance relationship with the marker.

While the marker variable has many advantages it also has some limitations. First is the marker variable is not related to the theoretical construct being considered and studied. Then the modification for CMV might be misleading and mistaken if the marker is by chance correlated with one or more key variables because of any unidentified share antecedent [33]. Moreover, there is a deep understanding of the constructs which are involved for the choice of marker variable and also require careful justification of theory. Thirdly, this technique is more effective at the detection of CMV than removing CMV. When method variance is not consistently distributed across measures or the marker variable is not properly selected, this technique may undervalue the extent of CMV [59].

In conclusion, in research for addressing common method bias, the marker variable technique is an effective, practical, and valuable tool. It is a strong tool because it is a simple conceptually as well as empirical. Also, it provides more precise inference about the relationships between constructs. On the other hand, its effectiveness and usefulness depend on its cautious choice and proper statistical analysis. With the passage of time, research methodologies advance, further modification and combination of marker variables' ability to enhance the credibility and legitimacy of empirical findings and results across various disciplines.

References

1. Wold, H.: Soft modeling: the basic design and some extensions. Systems Under Indirect Observation, Part II **2**, 36–37 (1982)
2. Hair, J.F., Black, W.C., Babin, B.J., Anderson, R.E.: Multivariate Data Analysis, Data Analysis, 8th edn. Cengage Learning, London (2018)
3. Hair, J.F., Hult, G.T.M., Ringle, C.M., Sarstedt, M.: A Primer on Partial Least Squares Structural Equation Modeling (PLS-SEM). Sage, Thousand Oaks, CA (2022)
4. Field, A.: Discovering statistics using SPSS for windows. Thousands Oaks, London (2000)
5. Pirouz, D.M.: An overview of partial least squares. Available at SSRN 1631359 (2006)
6. Abdi, H.: Partial least squares (PLS) regression. In: Lewis-Beck, M., Bryman, A., T. Futing, (eds.), Encyclopedia of Social Sciences Research Methods. Sage, Thousand Oaks (2003)
7. Dash, G., Paul, J.: CB-SEM vs PLS-SEM methods for research in social sciences and technology forecasting. Technol. Forecast. Soc. Chang. **173**, 121092 (2021)
8. Hair, J., Alamer, A.: Partial least squares structural equation modeling (PLS-SEM) in second language and education research: Guidelines using an applied example. Res. Methods Appl. Linguist. **1**(3), 100027 (2022)
9. Magno, F., Cassia, F., Ringle, C.M.: A brief review of partial least squares structural equation modeling (PLS-SEM) use in quality management studies. TQM J. **36**(5), 1242–1251 (2024)
10. Becker, J.M., Proksch, D., Ringle, C.M.: Revisiting Gaussian copulas to handle endogenous regressors. J. Acad. Mark. Sci. **50**(1), 46–66 (2022)
11. Ringle, C.M., Sarstedt, M., Sinkovics, N., Sinkovics, R.R.: A perspective on using partial least squares structural equation modelling in data articles. Data Brief **48**, 109074 (2023)
12. Chang, W., Franke, G.R., Lee, N.: Comparing reflective and formative measures: new insights from relevant simulations. J. Bus. Res. **69**(8), 3177–3185 (2016)
13. Markus, K.A., Borsboom, D.: Frontiers of Test Validity Theory: Measurement, Causation, and Meaning. Routledge (2024)
14. Howell, R.D., Breivik, E., Wilcox, J.B.: Reconsidering formative measurement. Psychol. Methods **12**(2), 205 (2007)
15. Coltman, T., Devinney, T.M., Midgley, D.F., Venaik, S.: Formative versus reflective measurement models: two applications of formative measurement. J. Bus. Res. **61**(12), 1250–1262 (2008)
16. Diamantopoulos, A., Siguaw, J.A.: Formative versus reflective indicators in organizational measure development: a comparison and empirical illustration. Br. J. Manag. **17**(4), 263–282 (2006)
17. Lee, N., Cadogan, J.W.: Problems with formative and higher-order reflective variables. J. Bus. Res. **66**(2), 242–247 (2013)
18. Borsboom, D., Mellenbergh, G.J., Van Heerden, J.: The theoretical status of latent variables. Psychol. Rev. **110**(2), 203 (2003)
19. Finn, A., Wang, L.: Formative vs. reflective measures: facets of variation. J. Bus. Res. **67**(1), 2821–2826 (2014)
20. Rose, J.M., Borriello, A., Pellegrini, A.: Formative versus reflective attitude measures: extending the hybrid choice model. J. Choice Modell. **48**, 100412 (2023)
21. Hanafiah, M.H.: Formative vs. reflective measurement model: guidelines for structural equation modeling research. Int. J. Anal. Appl. **18**(5), 876–889 (2020)
22. Bagozzi, R.P., Yi, Y.: On the evaluation of structural equation models. J. Acad. Mark. Sci. **16**, 74–94 (1988)

23. Hulland, J.: Use of partial least squares (PLS) in strategic management research: a review of four recent studies. Strateg. Manag. J. **20**(2), 195–204 (1999)
24. Chin, C.L., Yao, G.: Convergent validity. In: Encyclopedia of Quality of Life and Well-Being Research, pp. 1398–1399. Springer International Publishing, Cham (2024)
25. Zhu, W.: Which should it be called: convergent validity or discriminant validity? Res. Q. Exerc. Sport **71**(2), 190–194 (2000)
26. Kline, R.B.: Principles and Practice of Structural Equation Modeling. Guilford publications (2023)
27. Gold, A.H., Malhotra, A., Segars, A.H.: Knowledge management: an organizational capabilities perspective. J. Manag. Inf. Syst. **18**(1), 185–214 (2001)
28. Shrestha, N.: Detecting multicollinearity in regression analysis. Am. J. Appl. Math. Stat. **8**(2), 39–42 (2020)
29. Hair, J.F., Jr., Matthews, L.M., Matthews, R.L., Sarstedt, M.: PLS-SEM or CB-SEM: updated guidelines on which method to use. Int. J. Multivar. Data Anal. **1**(2), 107–123 (2017)
30. Kock, N.: Common method bias in PLS-SEM: A full collinearity assessment approach. Int. J. e-Collaboration (IJEC) **11**(4), 1–10 (2015)
31. Schwarz, A., Rizzuto, T., Carraher-Wolverton, C., Roldán, J.L., Barrera-Barrera, R.: Examining the impact and detection of the "urban legend" of common method bias. ACM SIGMIS Database: The DATABASE for Adv. Inf. Syst. **48**(1), 93–119 (2017)
32. Burton-Jones, A.: Minimizing method bias through programmatic research. MIS quarterly, 445–471 (2009)
33. Podsakoff, P.M., MacKenzie, S.B., Podsakoff, N.P.: Sources of method bias in social science research and recommendations on how to control it. Annu. Rev. Psychol. **63**(1), 539–569 (2012)
34. Kock, F., Berbekova, A., Assaf, A.G.: Understanding and managing the threat of common method bias: detection, prevention and control. Tour. Manage. **86**, 104330 (2021)
35. Jakobsen, M., Jensen, R.: Common method bias in public management studies. Int. Public Manag. J. **18**(1), 3–30 (2015)
36. MacKenzie, S.B., Podsakoff, P.M.: Common method bias in marketing: causes, mechanisms, and procedural remedies. J. Retail. **88**(4), 542–555 (2012)
37. Baumgartner, H., Weijters, B.: Commentary on "common method bias in marketing: causes, mechanisms, and procedural remedies." J. Retail. **88**(4), 563–566 (2012)
38. Chin, W.W., Thatcher, J.B., Wright, R.T.: Assessing common method bias: problems with the ULMC technique. MIS Q. 1003–1019 (2012)
39. Williams, L.J., Hartman, N., Cavazotte, F.: Method variance and marker variables: a review and comprehensive CFA marker technique. Organ. Res. Methods **13**(3), 477–514 (2010)
40. Chang, S.J., Van Witteloostuijn, A., Eden, L.: Common method variance in international business research. Res Methods Int. Bus. 385–398 (2020)
41. Podsakoff, P.M., MacKenzie, S.B., Lee, J.Y., Podsakoff, N.P.: Common method biases in behavioral research: a critical review of the literature and recommended remedies. J. Appl. Psychol. **88**(5), 879 (2003)
42. Tehseen, S., Ramayah, T., Sajilan, S.: Testing and controlling for common method variance: a review of available methods. J. Manage. Sci. **4**(2), 142–168 (2017)
43. Yüksel, A.: A critique of "Response Bias" in the tourism, travel and hospitality research. Tour. Manage. **59**, 376–384 (2017)
44. Liang, H., Saraf, N., Hu, Q., Xue, Y.: Assimilation of enterprise systems: the effect of institutional pressures and the mediating role of top management. MIS Q. 59–87 (2007)
45. Chang, S. E., Shen, W. C., Liu, A. Y.: Why mobile users trust smartphone social networking services? A PLS-SEM approach. J. Bus. Res. **69**(11), 4890–4895 (2016)
46. Abdi, H., Williams, L.J.: Principal component analysis. Wiley Interdisc. Rev. Comput. Stat. **2**(4), 433–459 (2010)
47. Fischer, D.G., Fick, C.: Measuring social desirability: short forms of the Marlowe-Crowne social desirability scale. Educ. Psychol. Measur. **53**(2), 417–424 (1993)
48. Ganster, D.C., Hennessey, H.W., Luthans, F.: Social desirability response effects: three alternative models. Acad. Manag. J. **26**(2), 321–331 (1983)

49. Crowne, D.P., Marlowe, D.: A new scale of social desirability independent of psychopathology. J. Consult. Psychol. **24**(4), 349 (1960)
50. Reynolds, W.M.: Development of reliable and valid short forms of the Marlowe-Crowne Social desirability scale. J. Clin. Psychol. **38**(1), 119–125 (1982)
51. Ramanaiah, N.V., Schill, T., Leung, L.S.: A test of the hypothesis about the two-dimensional nature of the Marlowe-Crowne social desirability scale. J. Res. Pers. **11**(2), 251–259 (1977)
52. Bagozzi, R.P., Yi, Y., Phillips, L.W.: Assessing construct validity in organizational research. Adm. Sci. Q. 421–458 (1991)
53. Kock, N., Lynn, G.S.: Lateral collinearity and misleading results in variance-based SEM: an illustration and recommendations. J. Assoc. Inf. Syst. **13**(7), 2 (2012)
54. de Andrade, L.G.M., Tedesco-Silva, H.: Recycling of predictors used to estimate glomerular filtration rate: insight into lateral collinearity. PLoS ONE **15**(2), e0228842 (2020)
55. Sterkin, A., Yehezkel, O., Bonneh, Y.S., Norcia, A., Polat, U.: Multi-component correlate for lateral collinear interactions in the human visual cortex. Vision. Res. **48**(15), 1641–1647 (2008)
56. Dormann, C.F., Elith, J., Bacher, S., Buchmann, C., Carl, G., Carré, G., Lautenbach, S., et al.: Collinearity: a review of methods to deal with it and a simulation study evaluating their performance. Ecography **36**(1), 27–46 (2013)
57. Lindell, M.K., Whitney, D.J.: Accounting for common method variance in cross-sectional research designs. J. Appl. Psychol. **86**(1), 114 (2001)
58. Leonard, S., Simmering, M.J., Wall, A.E.: A metacognitive approach to understanding what marker variables measure. Curr. Psychol. **43**(40), 31222–31240 (2024)
59. Richardson, H.A., Simmering, M.J., Sturman, M.C.: A tale of three perspectives: examining post hoc statistical techniques for detection and correction of common method variance. Organ. Res. Methods **12**(4), 762–800 (Accounting for Common Method Variance in Cross-Sectional Research Designs) (2009)

Chapter 2
PLS-SEM Path Model Estimation

Learning Objectives

1. Differentiate between measurement and structural models in PLS-SEM.
2. Evaluate model validity and reliability using confirmatory and exploratory factor analysis.
3. Analyze unobserved heterogeneity and apply mediation and moderation techniques.

2.1 PLS Path Model Assessment: Introduction

It is usual for researchers in business studies to use a structural model to determine the relationship between the study variables (usually the latent ones). For determining the statistical relationship, the model used must be reliable, valid, and representative of the proposed relationships between the variables [1, 2]. Generally, we speak about the measurement model and the structural model. The measurement model is aimed at measuring the relationship between the items and their representative construct. The structural model captures the relationship between the constructs or latent variables. PLS path model assessment is a technique to judge the suitability of the statistical models (both measurement and structural) [3, 4]. While applying PLS path modelling, researchers usually suppose that the data come from a homogeneous population, which is not realistic. Such an assumption generates misleading results and false interpretations [5]. Therefore, researchers must appreciate heterogeneity in the population for unambiguous results. Also, in many cases, researchers segregate the data based on the visible sample characteristics (age, gender, education, etc.) or through clustering (density-based clustering, k-means clustering, hierarchical clustering, etc.). However, both approaches do not work with SEM because it is hard to identify the heterogeneity [6]. In such circumstances, researchers must appreciate the methodological approach that uncovers the unobserved heterogeneity in the

X. Chew et al., *Partial Least Squares Structural Equation Modeling and Complementary Methods in Business Research*, Information Systems Engineering and Management 67, https://doi.org/10.1007/978-3-032-01055-1_2

data. PLS path modelling is a suitable method to assess the model's fitness, especially when there is unobserved heterogeneity in the data [7, 8]. This chapter aims to explore the issue of measurement model and structural model assessment primarily through confirmatory and exploratory factor analysis. The chapter also speaks about the issues of convergent validity and discriminant validity with the laid-down criteria for both validity measures. In the end, the discussion is made to understand the concept of moderation and mediation analysis, especially for structural model assessment.

2.2 Factor Analysis: Exploratory Versus Confirmatory Factor Analysis

Factor analysis includes a variety of statistical techniques used to identify the factors relevant to an abstract concept. The term factor refers to the unobservable characteristics of the individuals that can be determined through the other observable indicators. Brown [9] (p. 10) defined a factor as "an unobservable variable that influences more than one observed measure and that accounts for the correlations among these observed measures." Factor analysis is a technique that assists the operationalization of respondents' characteristics by developing the objective measures of these characteristics [10, 11]. It refers to a host of methods applied to develop an objective understanding of the latent variable from the underlying indicators (which are directly measurable) and also to affirm the theoretical explanation of the concept (latent variable). Cudeck [10] (p. 265) defined factor analysis as "a collection of methods for explaining the correlations among variables in terms of more fundamental entities called factors". The leading use of factor analysis is isolating indicators associated with a factor (latent variable, concept) from other indicators and determining the magnitude and strength of their relationship with the factor. The common use of factor analysis is to group similar variables into different groups consisting of homogeneous variables, where one group is distinct from the other group. The variables in the same group observed similar relational patterns, i.e., "arranging the like items with the like" [10] (p. 268). The major assumption of factor analysis is that the variables are related to each other to some extent. Factor analysis determines whether the different items converge to the same theoretical determining concept. There are two ways of doing factor analysis i.e., exploratory factor analysis (EFA) and confirmatory factor analysis (CFA). In the following, we explain the nature and intended use of the EFA and CFA. Our explanation of both is aimed at facilitating the novice researcher to understand the basic differences between both and the purpose of using both.

2.2.1 Exploratory Factor Analysis (EFA)

As discussed earlier that social science research is dominated by quantitative studies, which require the development of measurement models to gauge theoretical concepts to be used in the study. It is always challenging for researchers, especially novice ones, to develop measurement models because the reliability of the findings is entirely dependent on the accuracy of measurement models [10, 12]. Once the constructs are measured, they are incorporated into the questionnaire to obtain the responses using different scales. For the development and construction of a questionnaire and construction of questionnaire items, it requires researchers to conduct factor analysis to ensure that the latent variables developed through factor analysis truly represent the theoretical concept [13]. One way of doing factor analysis is 'Exploratory Factor Analysis' (EFA). When a latent variable is represented by many indicators, EFA is used to reduce the number of indicators that best represent the latent variable. The main purpose of EFA is to unveil the relational pattern[s] in the original data. It explains how and in how many ways the indicators and the constructs are interrelated. In this way, it helps in the development of new theories [12, 14]. Usually, EFA is used at the initial stages of data collection, particularly at the time of questionnaire development. EFA helps identify the unrelated items (to a concept/factor/latent variable) that must not be included in the questionnaire. It is also used to explain the survey instrument (questionnaire) dimensionality [10]. It must be noted that EFA should not be confused with principal component analysis (PCA). They both are different across different dimensions. Though PCA is also used to reduce large data into smaller one, keeping intact the significant relational pattern in the data [15]. Principally, in EFA, the underlying variables are defined by the latent variable (factor) and any change in the latent will influence the relational pattern i.e. the latent variable defines the observed relationship for the observed items. Whereas in PCA, purpose is dimensionality reduction through the identification of uncorrelated items called principal components [12, 16]. Similarly, EFA assumes that latent variables are not directly measurable, whereas PCA supposes that observable variables have linear relationships and the variance in the data is dependent on the linear relationship between the observed variables [15]. EFA also emphasizes on the 'latent construct i.e. latent construct defines the items, whereas PCA does not consider latent construct to reduce the data. EFA can be performed using two statistical techniques i.e. Q-type factor analysis and R-type factor analysis. If the factors are coming from the respondents, then Q-type test is recommended and if the factors are deduced from correlational matrix, then R-type test is preferred. EFA is carried out multiple times to reduce the factors to develop a hypothetical model representing the derivation of latent variables from the underlying items [10, 12].

2.2.2 *Confirmatory Factor Analysis (CFA)*

As depicted by its name, CFA is utilized to confirm the existing theoretical models. CFA confirms whether the collected data supports the hypothesized relationships or not. Once the hypothetical model is developed then it can be tested through confirmatory factor analysis (CFA). In simple terms, CFA is confirmation of EFA. CFA is a statistical procedure used to confirm the alignment between the observed measurements and the researcher's a priori conceptualization, a fit between the measurement of the concept and its theoretical explanation [17, 18]. Researchers need to meet some a priori criteria before using CFA. For example, the latent construct is theoretically saturated, and the survey instrument has already been confirmed for internal validity. Once it is done, the relationship must be specified between the indicators and the latent construct. After that, FCA is used to obtain the number of latent constructs, to determine the relational pattern within the constructs and constructs and their respective items [19, 20]. For novice researchers, it is recommended to use FCA when they adopt the existing and established survey instrument (questionnaire for a similar sample (in terms of relevant characteristics of the population and sample. There are a variety of techniques to perform CFA, such as maximum likelihood estimation, weighted least square, multi-group CFA, second-order CFA, Monte Carlo Simulation, etc. Researchers have the liberty to use specific techniques [s] as deemed fit. CFA confirms whether the hypothesized relationships are supported by the statistical testing of the data [17, 21].

Ideally, the results obtained through EFA must be confirmed with CFA before actually using the measurement instrument for data collection. However, EFA and CFA must be carried out using different representative samples. If EFA and CFA are performed with the same sample, then CFA will almost show the same relationship as demonstrated by EFA and will lead to a generalizability issue. Also, there is a possibility that when using a different sample for EFA and CFA, the new sample may respond to certain items entirely differently from the earlier sample. Hence, it is recommended to use two different samples, both for EFA and CFA [21, 22]. In case of a relatively larger sample, that can be randomly divided into two sample sets. It always important to perform EFA and CFA before full-scale utilization of the survey instrument, otherwise, there are chances that the scale will not fit the theoretical explanation of the concept [23].

2.2.3 *Procedure for Doing Factor Analysis (EFA and CFA)*

Doing factor analysis involves a variety of considerations on the part of the researcher, and the researcher must use a holistic approach, keeping in view the context of the study and the theoretical underpinning of the concept [24, 25]. Factor analysis is based on the available information that helps researchers to build an argument about the construction of a measurement tool. Researchers must spend sufficient time collecting the relevant theoretical and contextual information for conceptual saturation before developing and employing the survey instrument. The researcher should not only rely on the a priori held personal assumptions about the relational pattern between the individual items and the underlying factor, but should focus more on a good collection of theoretical and empirical evidence for a well-constructed survey instrument [25, 26].

The first step in doing the factor analysis is to adequately define the construct and its intended purpose. This is essentially a prerequisite to define the purpose for which the instrument is to be used. Similarly, the latent factor or the theoretical concept that is intended to be measured must also be clearly defined in alignment with the widely understood explanation of the concept. Once the purpose of using a measurement instrument and the factor are fully defined and understood, the next step is to ensure that the items chosen to represent a factor truly represent the factor [27]. Knekta et al. [1] suggested that while performing CFA, researchers must ensure that the instrument contains a minimum of four items. The second step is to select an appropriate sample size to do the factor analysis. Generally, a large sample size is preferred because it provides more statistical accuracy. Some of the scholars argue that a sample size of 100 to 200. Early guidelines on sample sizes for factor analysis were general in their nature, such as a minimum sample size of 100 or 200 (e.g., see [28–30]). Although it is very tempting to adopt such general guidelines, caution must be taken, as they might lead to underestimating or overestimating the sample size needed [31, 32].

2.3 Convergent Validity

Researchers always strive to obtain results that must be widely recognised and accepted by the research community. Robustness of the research findings is the single most important criteria to judge research quality. There are a variety of subjective (qualitative) and objective (quantitative) ways used to attain the robustness of the findings. Researchers must be careful in their choice of sample selection, variable measurement, and statistical techniques to ensure that the findings correspond to the methodological choice. For instance, while developing the measure of a latent concept, researchers must ensure that all the dimensions and components have been captured in defining the concept, and the items selected truly define the concept [33]. This means that using different measures to capture the concept will be closely relevant and ensure the convergence to the common construct. Convergent validity is one of the most common techniques to ensure the convergence of different measures to a common construct. In simple words, convergent validity defines how closely a new measure of a construct is aligned with the already established measures. It becomes extremely important to ensure the convergent validity of a measure (measurement instrument/questionnaire) when the measure is adapted with the incorporation of the new dimensions into the measurement instrument [34, 35]. A high convergent validity denotes that different measures used to capture the same construct generate similar results. Carlson and Herdman [36] (p. 17) defined convergent validity as "the extent to which two measures capture a common construct". We try to define the concept with an example. If we use two measures to determine self-confidence and self-efficacy, then it is expected that individuals reporting high score of self-confidence are likely to report high score on self-efficacy, hence the two measures are highly correlated and thus the existence of convergent validity is established. If the two measures are not correlated and less convergent, then it leads to ambiguity in the interpretation of the findings and the findings are not considered to be believable. It is critically important to establish the convergent validity of the construct/survey instrument because it significantly influences the study's findings [34, 37].

The correlation coefficient is the common measure of convergent validity. A value of $r = 0$ means there is no convergent validity between the two constructs, whereas $r = 1$ indicates perfect convergence or convergent validity. The value of 'r' can range between 0 and 1, i.e., from no convergent to 100 percent convergence [36, 38]. For instance, Campbell and Fisks [39] found 144 different values of 'r' ranging from 0.02 to 0.82. However, what the 'r' value defines as the high correlation between the construct is not consented between the scholars [36]. For example, for some researchers, $r = 0.28$ shows enough convergence (e.g., Larraza-Kintana et al. [40]), and for others (e.g., Podsakoff et al. [41]), $r = 0.75$ does not show a sufficient level of convergence. Similarly, Wall et al. [42] found that the value of 'r' can range from 0.26 to 0.65. Broadly, among the researchers, convergent validity reflects the extent to which the items of an underlying construct are internally consistent with each other [43]. Historically, the reliability of the measurement scale has been used to check the convergent validity [44]. However, some researchers argue that scale reliability is not an adequate and appropriate judgment of convergent validity. Another measurement is CFA, which is more inclined towards determining the model's fitness [44, 45]. However, due to the issue of model measurement quality, CFA also does not deem to give an accurate measure of convergent validity. Therefore, Cheung et al. [46] proposed some additional criteria to ensure the convergent validity of the measurement. The first one is that the value of standardized factor loading should be at least equal or greater than 0.5, and ideally it should be close to 0.7 or above. The other criterion is average extracted variance value (AEV) should be higher than 0.5. Kock [47, 48] suggested that a measure or measurement instrument is considered to reflect acceptable convergent validity if the questionnaire statements are understood by the respondents in the same way as they are intended to be understood by the researcher. It is a link between the construct indicators and the construct.

Researchers use proxies to develop the measure of a latent construct. The proxies are observable measures of the unobservable constructs [49, 50]. Proxies are also used when a construct is not correctly defined or there are some restrictions to direct access to the data pertinent to the desired construct. The choice of appropriate proxies is dependent on the refinement of the research question and the selection of appropriate theory[ies] that theoretically define the relationship between the construct and its proxies [51]. For example, there could be different proxies (product record, task completion, supervisor rating, etc.) for measuring the employees' performance, the choice is dependent on the availability and suitability of the proxy to represent the construct 'employee performance'. These proxies are important to establish the convergent validity. Researchers are more interested in knowing the existing of a higher correlation (strong convergent validity) exists between these proxies to be used as alternative measures of the construct. A strong convergent validity is much appreciated, i.e., both proxies are necessary to generate the same result (e.g., employee performance), i.e. the two measures report the same value for r with the other variable (e.g., customer satisfaction). A deviation of r value from $r = 1$ signals that the proxy is not a true measure of the construct. Similarly, a value of $r = 0$ means that the proxy is independent of the construct, and $r = 1$ means that the proxy is a perfect substitute for the construct [45, 46].

In the context of social and behavioural studies, the constructs used are not directly measurable. In such circumstances, researchers have a choice either to use the established measures from the previous studies or to use the revised scale. In some cases, it becomes imperative to translate the scale or revise the scale that suit the respondents. Researchers must be careful in such cases and must perform pre-testing of the revised measure before full-scale utilization. Cronbach's alpha and confirmatory factor analysis (CFA) are measures to check the quality of the measurement scale's correspondence to the underlying latent construct. Cheung et al. [46] suggested that in some cases use of Cronbach's alpha as a measure of convergent can be misleading when the researchers overlook sampling error. Sampling error exists when the sample does not truly represent the population. The common measure of sampling error is standard sampling error, i.e., variability of mean or standard deviation across different samples [36].

2.3.1 Factor Loading

Spearman's [52] seminal work about the general intelligence and its relationship with the cognitive abilities laid down the foundation of factor analysis and its applications across different fields. Spearman explained how and what cognitive abilities are related to human intelligence [53]. This relationship was later known as factor loading. Factor loading is critical to understanding how well the observable items/indicators are related to a latent variable, subject to further analysis. In many instances, constructs are measured through observable variables (items) [54]. Ideally, all the underlying variables of a construct must be perfectly correlated with the construct. Factor loading is a statistical technique that determines the extent of the relationship between a construct and its defining variables, i.e., the correlation between the two [55]. Factor loading is critical to establish the validity and reliability of the survey instrument. Factor loading determines the extent and strength of the relationship between the items and their representative construct determined through the coefficient of correlation 'r' and its value ranges from -1 to $+1$, the sign shows the direction of the relationship, and the digit shows the strength of the relationship [56, 57]. Factor loading is an essential component of factor analysis. Factor loading primarily serves three purposes. First, to establish the validity of the construct (i.e., confirms if the items truly represent the construct). Second, it helps to locate the patterns and structure in the data, and third, it aids in the refinement of the survey instrument. The researcher excludes the items that do not load on the factor/construct [58]. While doing the factor loading researcher should take care of the issues related to the choice of appropriate statistical methods, sampling issues, and transparency in reporting the factor loading. Any manipulation of the data or the sample may result in statistically ambiguous and vague results [56, 57].

2.3.2 *Average Variance Extracted (AVE)*

Average Variance Extracted (AVE) is a statistical measure to be used in factor analysis and structural model assessment to determine the convergent validity of the construct/ latent variable [44]. It refers to the variance in items explained by the construct and indicates if the items well represent a construct. AVE is generally used to assess convergent validity and construct reliability [59]. This must be noted that AVE is not a measure of discriminant validity. To calculate AVE, the first step is to know the value of factor loading for each item and then to square those values. Below is formula to calculate AVE.

$$AVE = \frac{\text{sum of the squared factor loading of items}}{\text{No. of items representing the construct}}$$

As a general rule value of AVE > 0.5 is acceptable, i.e. construct explains more than 50% variance in the items that show a higher alignment between the construct and its items [60, 61].

2.3.3 *Composite Reliability*

Composite reliability is a statistical measure of the internal consistency between the items that measure the same construct [60]. In other words, composite reliability shows the strength of the relationship between the items related to a single construct. Factor loading is a common method used to determine the composite reliability [62]. While calculating the composite reliability, the first step is to identify the items that are used to measure the construct. After that, factor loading is performed using the coefficient of correlation. The next step is to calculate error variance (1-AVE) for every item. Error variance is the variance in the individual item not explained by the construct. The last step is to apply the formula for composite reliability, given below

$$CR = \frac{\text{Sum of the squared loading of all items}}{\text{Sum of the squared factor loading of all items} + \text{Sum of error variance of all items}}$$

As a general rule, a value of CR > 0.7 is considered acceptable, reflecting an appropriate level of internal consistency between items [45, 63]. There are a few assumptions of composite reliability. First is the one-dimensionality, i.e., the items should measure a single construct. Second, there should be a linear relationship between the items and the construct. Third, no cross-loading, and fourth, the sample size is right. Many of the researchers suggest a sample size of minimum 300 should be enough for getting better reliability [64, 65].

2.3.4 Cronbach's Alpha

Lee [66] introduced a statistical technique (later named Cronbach's alpha) to determine the internal consistency between the items grouped on a measurement scale. It is also used to judge the reliability of a measurement scale. Generally, a value of Cronbach's alpha higher than 0.7 is acceptable, and the scale is considered reliable. However, the researchers like Carmines and Zeller [67], Nunnally [68], and Lance et al. [69] suggested that a value of 0.8 represents an appropriate level of scale reliability, and 0.7 is a modest level of scale reliability. Cheung et al. [46] discussed that the Cronbach's alpha assumption of equal factor loading does not hold in the case of latent constructs. If a latent construct is represented by the items that exhibit unequal factor loading, then Cronbach's alpha underrates the reliability of the scale [70].

2.4 Discriminant Validity

Discriminant validity is used to establish construct validity. We have defined the concept of discriminant validity in the previous chapter. Just to remind the reader, we reproduce that discriminant validity confirms that a test measures what it intends to measure. It is also called divergent validity. Discriminant validity only accounts for the difference between the tests used to determine different construct [46]. In the following, we describe the various methods used to determine the discriminant validity of a test.

2.4.1 Fornell-Larcker Criterion

Fornell and Larcker [44] introduced Fornell-Larcker criteria to be used in SEM analysis for the assessment of the discriminant validity. It measures how distant a construct is from the other constructs in a statistical model. While using the criterion, the square root of the value of AVE is compared to correlations between the latent variables. The construct is valid and distinct if the value of the square root of AVE of a construct is greater than the highest value of the construct's correlation with other constructs [46]. The following steps are followed to determine the discriminant validity through the Fornell-Larcker criterion.

(i) Calculate AVE value for each construct by taking the square of the factor loading (correlation coefficient of each observable indicator with the latent construct) for all the indicators, adding them all, and dividing the resultant value by the total number of indicators of a latent construct.
(ii) Take the square root of the AVE value as calculated in step 1
(iii) Develop a table to exhibit the correlations between all the construct (Correlation matrix)
(iv) For every construct, compare the value in step (ii) with the all the values in step (iii)
(v) The value obtained in step (ii) must be greater than the highest correlation coefficient value among all the values in step (iii). If this is the case, then the discriminant validity of the construct is established via Fornell-Larcker criterion i.e. the construct is inherently different form all the other constructs used in the model.

However, the above-mentioned steps are only meant to provide some explanation about the Fornell-Larcker criterion. Researchers do not need to manually do this all. There are available software like Smart PLS, Wrap PLS, SPSS. Stata, etc. that include the Fornell-Larcker criterion assessment. Though Fornell-Larcker criterion remained in use (still some use it) for long as a reliable measure of discriminant reliability, but some recent studies have shown that it should be no more used and should be discontinued as an assessment of discriminant validity. The criterion does not work when there is slight difference in the factor loadings on the same construct ($0.65 \geq \lambda \leq 0.85$) [71]. An alternate and currently widely used assessment of discriminant validity is Heterotrait-Monotrait ratio (HTMT), which we explain below.

2.4.2 *Heterotrait-Monotrait Ratio (HTMT)*

In 2015 was introduced by Henseler, Ringle and Sarstedt introduced an advanced technique called Heterotrait-Monotrait ratio of correlations (HTMT) to assess the discriminant validity in SEM analysis. HTMT determines that two theoretically different concepts are also different statistically. It is one of the widely used criteria in business studies. It provides an estimate of the correlation between two [latent] variables. A value of HTMT 0.90 confirm the existence of discriminant validity, whereas a value > suggests a lack of discriminant validity i.e. the construct is not theoretically distinct. Hair et al. [56] (p. 79) defined HTMT as ratio between the "the mean value of the indicator correlations across constructs (i.e., the heterotrait–heteromethod correlations) relative to the (geometric) mean of the average correlations for the indicators measuring the same construct (i.e., the monotrait–heteromethod correlations). In the following, we try to explain the HTMT in simple steps.

(i) Determine the correlations between the indicators of different constructs in the model and calculate the mean value (heterotrait–heteromethod correlations).
(ii) Repeat the same for the correlations between the indicators for a single construct (monotrait–heteromethod correlations).
(iii) Calculate the ratio between the mean value (step-i) and the mean value (step-ii).
(iv) Ratio <0.90 confirms the discriminant validity of the construct, i.e. construct is different from the other constructs in the model. Ideally, the value should be <0.9 for the conceptually similar constructs, and for the constructs theoretically different the value of HTMT < 0.85 [56].

2.4.3 *Cross Loading*

Cross-loading happens when one item loads (correlated with) two or more two factors. Correlation coefficient (r) is a usual measure of factor loading. In many cases, one item should be highly correlated to its own construct in comparison to it correlation with the other construct of the study [56, 57]. The value of 'r'> is higher than threshold value of 0.3 across two or more factors reflects significant level of cross loading. Presence of cross loading is problematic because it makes difficult to clearly identify and interpret the relationship between the items and constructs [72]. A common solution to the problem is to exclude the cross-loading items from the analysis for clear and more reflective relationship between the construct and underlying item. Schreiber [26] (p. 1009) argues that the choice of retaining an item is "generally arbitrary" and there are no available criteria to judge the level of significant cross loading. It depends on variety of underlying reasons. A loading of 0.40 could be higher on one occasion for a researcher but it could be smaller for the same on some other occasion. The selection of threshold level must be supported by a logical and rationale argument. However, in any case the issue of cross loading must not be overlooked, and the researcher must explain it theoretically.

2.5 Moderation and Mediation

In business and other social sciences research, the moderating variable has a notable role in explaining the relationship between the predictor and outcome variables. A moderating variable is understood to strengthen or weaken the relationship between the two variables and influences the nature of their relationship [73]. In business research, it is commonly found that the relationship between two variables is dependent on a third variable. For instance, the influence of the workplace environment on the employees' job performance is dependent on the workplace relationship. Similarly, goal-setting theory [74] argues that setting high work expectations influences the work performance of different employees differently, depending on their respective levels of ability to perform difficult tasks. Traditionally, business and other social science research is dominated by quantitative studies, and many of such studies include the role of a moderator to understand the proposed relationships deeply [75]. In pure research language, a moderator is a variable that influences the nature and magnitude of the relationship between two variables, whereas moderation refers to the extent or degree of effect on such a relationship [73]. There could be different levels of moderation, like simple, partial, and full moderation [76]. A simple level of moderation exists when a moderator influences the relationship between the two variables. For instance, the linear relationship between an independent variable (A) and a dependent variable (B) can be stated through the following simple regression equation:

$$B = \beta_0 + \beta_1 A + \varepsilon \tag{2.1}$$

Above, β_0 is the intercept, i.e., expected value of B when A = 0, β_1 coefficient of A, and denotes the change in B because of one point change in A, and ε is the error term. When we introduce the moderator 'C' in the above relationship, then this is expressed by the following regression equation.

$$B = \beta_0 + \beta_1 A + \beta_2 C + \beta_3 (A \times C) + \varepsilon \tag{2.2}$$

The above equation includes both the moderator 'C' and the interaction term (A × C), called a two-way interaction that involves the independent variable and the moderating variable. In simple words, two-way interaction is the combined effect of two variables (independent and moderating) on another variable (dependent). Analysis of Variance (ANOVA) is a common statistical measure of two-way interaction. It determines the joint effect of two independent variables (the moderator is treated as independent) on a dependent variable [73, 77]. The interaction term is central to determining the significance level of the moderator. It shows if the relationship between independent and dependent variables changes with the change in the value of the moderator. Where β_1 and β_2 determine the main effect, β_3 confirms the existence of a moderation effect. The inclusion of the main effects is mandatory in the equation; otherwise, the results cannot be interpreted properly. For the general understanding of the concept, we include only one moderator, but in some other cases, the relationship between the independent and the dependent variables can be influenced by more than one moderator or multiple moderators. In that case, we need to include the main effect for each of the moderators and the interaction term with each moderator (between the independent and moderators) and the interaction term between the independent and all the moderators. For example, with the inclusion of another moderator 'E' in the above example, the regression equation will be depicted as under.

$$\begin{aligned} B = {} & \beta_0 + \beta_1 A + \beta_2 C + \beta_3 E + \beta_4 (A \times C) + \beta_5 (A \times E) \\ & + \beta_6 (C \times E) + \beta_7 (A \times C \times E) + \varepsilon \end{aligned} \tag{2.3}$$

The above equation shows the main effects of one independent and two moderating variables, three two-way interaction terms, and one three-way interaction term. The three-way interaction term determines the combined effect of three variables (one independent and two moderators) on the dependent variable. The three-way interaction specifies that if the moderating effect of the first moderator on the relationship between the independent variable and the dependent variable is itself moderated by the inclusion of the second moderator.

Coming back to the types of moderation, we already defined the concept of simple moderation. The other type of moderation is partial moderation. Partial moderation exists when the relationship between the independent variable and the dependent variable still exists even if the moderator is not included. For instance, if we consider the influence of transformative leadership on employees' motivation, with the leader's gender as a moderator. The relationship between the two will still exist even if the leader's gender is not considered. We can say that a leader's gender partially moderates the relationship between transformative leadership and employees' motivation. Full moderation exists when the relationship between the other two variables (independent and dependent) is completely dependent on the third variable (moderator), and the influence of the independent variable on the dependent becomes zero for the specific value of the moderator. In other words, the influence of the independent variable on the dependent variable changes with the change in the value of the moderator, and for a certain value of the moderator, it becomes zero [73, 78].

Some important questions must be considered while specifying the relationship between the variables. For instance, with a simple regression involving two variables and one moderator, it is identical to consider either of the independent variables or the moderator as an independent variable. For instance, as shown in Eq. 2, either 'A' or 'C' can be accounted as an independent variable. However, this is entirely dependent on the theoretical understanding of the research problem on the part of the researchers, i.e., how they deduce these relationships from the existing literature. Another important issue is the categorical nature of the moderator. For instance, as we mentioned earlier that the relationship between transformative leadership and employees' motivation is moderated by gender. In this case, the gender is categorized into two categories or has binary categories. In that instance, coding the value as 0, 1 for the category makes the interpretation easier, while the testing moderating influence follows the same procedure as for the single category moderator. But in case the moderator has more than two categories, then testing the moderating influence is a little difficult. It can be done using either Analysis of Co-variance (ANCOVA) or through simple regression and using dummy variables. Dummy variables are used for all the categories of the moderator except one, which is known as the reference category. It is used as a baseline to compare the moderation results with the other categories [73, 79].

2.5.1 What is Mediation

One of the fundamental purposes of [social science and business] research is to understand the underlying processes [es] of an observable social phenomenon. For a reputable study, it may not be sufficient to describe and test the relationship between the independent and dependent variables, but it needs to understand these relationships beyond that. This is especially in the case in complex social and business phenomena, where there are several contributing factors that make a phenomenon happen. Researchers can better explain such complex phenomena by considering mediation analysis. It is considered an important type of analysis that helps to unveil the relationships that cannot be understood by merely considering them in terms of independent and dependent variables [80]. In simple words, a mediator variable works as an intermediary variable between an independent variable and a dependent variable. The relationship between an independent variable and a dependent variable exists only in the presence of the mediator. It is an intervening variable that transmits the effect of the independent variable on the dependent variable [81]. In other words, it explains the mechanism of the relationship between two related variables. To be more precise, a change in independent variable will cause the change in mediator that in turn results into change in the value of the dependent variable. If such type of relationship between the independent variable and dependent variable exists, then researchers perform the mediation analysis [80].

Mediation analysis "is now almost mandatory for new social-psychology manuscripts." [82], (p. 550). Fiedler et al. [80] (p. 1231) defined mediation analysis as a "*a statistical procedure to test whether the effect of an independent variable X on a dependent variable Y (i.e., X → Y) is at least partly explained by a chain of effects of the independent variable on an intervening mediator variable Z and of the intervening variable on the dependent variable (i.e., X → Z → Y)*". The mediating effect can also be understood in terms of direct and indirect effects. Direct effect exhibits the direct causal relationship between two (latent) variables and is presented by a single arrow with direction from cause[variable] to effect[variable] as described above. The indirect effects [s] are represented by structural paths describing the sequential relationship with at least one mediator. The paths are presented by several arrows, depending on the path. For example, in the above definition, there are two paths with two arrows showing the sequential relationship. Hair et al. [57] (p. 142) described five types of mediation. These include 'complementary mediation' (both direct and indirect effect are statistically significant [p-value < 0.05] and follow the same direction), competitive mediation (both effects are significant but in opposite direction), indirect mediation (only the indirect effect is significant), direct-only non-mediation (only direct effect is significant), and non-effect-non-mediation (both effects are insignificant). In last two cases, the mediation does not exist between the independent variable and the dependent variable. Zhao et al. [83] and Kenny and Baron's [84] called complimentary mediation as partial mediation and indirect only mediation as full mediation. Partial mediation exists when the mediating variable explains only a few [or one] but not all the relationship between independent variable and dependent variable. In other words, in case of partial mediation, direct relationships may also exist between the independent and dependent variable in addition to the direct relationship between the mediator and dependent variable. Whereas, for full mediation, the mediating variable captures all the relationships between the two variables, i.e. independent variable and dependent variable are only related through mediation, and if the mediator is excluded, then no such relationship exists [80, 83]. A structural model may include one mediator or more than one mediator, called multiple mediators. In the case of one mediator, single mediation analysis is performed, while in the case of multiple mediators, researchers perform multiple mediation analysis. Multiple mediation includes both direct and indirect effects, and multiple mediation analysis counts for the sum of direct and indirect effects (see Hair et al. [57], p. 144).

References

1. Knekta, E., Runyon, C., Eddy, S.: One size doesn't fit all: using factor analysis to gather validity evidence when using surveys in your research. CBE—Life Sci. Educ. **18**(1), rm1 (2019)
2. Vinzi, V.E., Trinchera, L., Amato, S.: PLS path modeling: from foundations to recent developments and open issues for model assessment and improvement. In: Handbook of Partial Least Squares: Concepts, Methods And Applications, in 47–82 (2009)
3. Hancock, G.R., Mueller, R.O.: The reliability paradox in assessing structural relations within covariance structure models. Educ. Psychol. Measur. **71**(2), 306–324 (2011)
4. Rigdon, E.E.: Rethinking partial least squares path modeling: in praise of simple methods. Long Range Plan. **45**(5–6), 341–358 (2012)
5. Van Bruggen, V., Ten Klooster, P.M., Van der Aa, N., Smith, A.J., Westerhof, G.J., Glas, G.: Structural validity of the world assumption scale. J. Trauma. Stress **31**(6), 816–825 (2018)
6. Bécue-Bertaut, M., Pagès, J.: Multiple factor analysis and clustering of a mixture of quantitative, categorical and frequency data. Comput. Stat. Data Anal. **52**(6), 3255–3268 (2008)
7. Rigdon, E.E.: Choosing PLS path modeling as analytical method in European management research: a realist perspective. Eur. Manag. J. **34**(6), 598–605 (2016)
8. Rigdon, E.E., Becker, J.M., Sarstedt, M.: Parceling cannot reduce factor indeterminacy in factor analysis: a research note. Psychometrika **84**(3), 772–780 (2019)
9. Brown, T.A.: Confirmatory Factor Analysis for Applied Research. Guilford publications (2015)
10. Cudeck, R.: Exploratory factor analysis. In: Handbook of Applied Multivariate Statistics and Mathematical Modeling, pp. 265–296. Academic Press (2000)
11. Yong, A.G., Pearce, S.: A beginner's guide to factor analysis: focusing on exploratory factor analysis. Tutorials in Quantitative Methods Psychol. **9**(2), 79–94 (2013)
12. Watkins, M.W.: Exploratory factor analysis: a guide to best practice. J. Black Psychol. **44**(3), 219–246 (2018)
13. Fabrigar, L.R., Wegener, D.T., MacCallum, R.C., Strahan, E.J.: Evaluating the use of exploratory factor analysis in psychological research. Psychol. Methods **4**(3), 272 (1999)
14. Beavers, A.S., Lounsbury, J.W., Richards, J.K., Huck, S.W., Skolits, G.J., Esquivel, S.L.: Practical considerations for using exploratory factor analysis in educational research. Pract. Assess. Res. Eval. **18**(6), n6 (2013)
15. Kherif, F., Latypova, A.: Principal component analysis. In Machine Learning, pp. 209–225. Academic Press (2020)
16. Sarkar, S.K.: Principal component analysis. In: Statistical Procedures for Analyzing Agricultural Data Using R, p. 139 (2023)
17. Hox, J.J.: Confirmatory factor analysis. Encycl. Res. Methods Criminol. Crim. Justice **2**, 830–832 (2021)

18. Kline, R.: Exploratory and confirmatory factor analysis. In: Applied Quantitative Analysis in Education and the Social Sciences, pp. 171–207. Routledge (2013)
19. Gatignon, H.: Confirmatory factor analysis. In: Statistical Analysis of Management Data, pp. 77–154. Springer, US, Boston, MA (2013)
20. Schmitt, T.A.: Current methodological considerations in exploratory and confirmatory factor analysis. J. Psychoeduc. Assess. **29**(4), 304–321 (2011)
21. Goretzko, D., Siemund, K., Sterner, P.: Evaluating model fit of measurement models in confirmatory factor analysis. Educ. Psychol. Measur. **84**(1), 123–144 (2024)
22. Aljaberi, M.A., Lee, K.H., Alareqe, N.A., Qasem, M.A., Alsalahi, A., Abdallah, A.M., Lin, C.Y., etal.: Rasch modeling and multilevel confirmatory factor analysis for the usability of the impact of event scale-revised (IES-R) during the COVID-19 pandemic. In: Healthcare, vol. 10, No. 10, p. 1858. MDPI (2022)
23. Bandalos, D.L., Finney, S.J., Hancock, G.R., Mueller, R.O.: The reviewer's guide to quantitative methods in the social sciences (2010)
24. Shrestha, N.: Detecting multicollinearity in regression analysis. Am. J. Appl. Math. Stat. **8**(2), 39–42 (2020)
25. Shrestha, N.: Factor analysis as a tool for survey analysis. Am. J. Appl. Math. Stat. **9**(1), 4–11 (2021)
26. Schreiber, J.B.: Issues and recommendations for exploratory factor analysis and principal component analysis. Res. Social Adm. Pharm. **17**(5), 1004–1011 (2021)
27. Williams, B., Onsman, A., Brown, T.: Exploratory factor analysis: a five-step guide for novices. Australasian J. Paramed. **8**, 1–13 (2010)
28. Boomsma, A.: Reporting analyses of covariance structures. Struct. Equ. Model. **7**(3), 461–483 (2000)
29. Comrey, A.L., Lee, H.B.: A First Course in Factor Analysis. Psychology press (2013)
30. Gorsuch, R.L.: Exploratory factor analysis. In: Handbook of Multivariate Experimental Psychology, pp. 231–258. Springer, US, Boston, MA (1988)
31. Wolf, E.J., Harrington, K.M., Clark, S.L., Miller, M.W.: Sample size requirements for structural equation models: an evaluation of power, bias, and solution propriety. Educ. Psychol. Measur. **73**(6), 913–934 (2013)
32. Worthington, R.L., Whittaker, T.A.: Scale development research: a content analysis and recommendations for best practices. Couns. Psychol. **34**(6), 806–838 (2006)
33. Higgins, W.C., Kaplan, D.M., Deschrijver, E., Ross, R.M.: Construct validity evidence reporting practices for the reading the mind in the eyes test: a systematic scoping review. Clin. Psychol. Rev. **108**, 102378 (2024)
34. Kaul, S., Boyle, K.J., Kuminoff, N.V., Parmeter, C.F., Pope, J.C.: What can we learn from benefit transfer errors? Evidence from 20 years of research on convergent validity. J. Environ. Econ. Manag. **66**(1), 90–104 (2013)
35. Zhu, W.: Which should it be called: convergent validity or discriminant validity? Res. Q. Exerc. Sport **71**(2), 190–194 (2000)
36. Carlson, K.D., Herdman, A.O.: Understanding the impact of convergent validity on research results. Organ. Res. Methods **15**(1), 17–32 (2012)
37. Chin, C.L., Yao, G.: Convergent validity. In: Encyclopedia of Quality of Life and Well-Being Research, pp. 1398–1399. Springer International Publishing, Cham (2024)
38. Cowin, L.S., Hengstberger-Sims, C., Eagar, S.C., Gregory, L., Andrew, S., Rolley, J.: Competency measurements: testing convergent validity for two measures. J. Adv. Nurs. **64**(3), 272–277 (2008)
39. Campbell, D.T., Fiske, D.W.: Convergent and discriminant validation by the multitrait-multimethod matrix. Psychol. Bull. **56**(2), 81 (1959)
40. Larraza-Kintana, M., Wiseman, R.M., Gomez-Mejia, L.R., Welbourne, T.M.: Disentangling compensation and employment risks using the behavioral agency model. Strateg. Manag. J. **28**(10), 1001–1019 (2007)
41. Podsakoff, N.P., Whiting, S.W., Podsakoff, P.M., Blume, B.D.: Individual-and organizational-level consequences of organizational citizenship behaviors: A meta-analysis. J. Appl. Psychol. **94**(1), 122 (2009)

42. Wall, T.D., Michie, J., Patterson, M., Wood, S.J., Sheehan, M., Clegg, C.W., West, M.: On the validity of subjective measures of company performance. Pers. Psychol. **57**(1), 95–118 (2004)
43. Bagozzi, R.P.: An examination of the validity of two models of attitude. Multivar. Behav. Res. **16**(3), 323–359 (1981)
44. Fornell, C., Larcker, D.F.: Evaluating structural equation models with unobservable variables and measurement error. J. Mark. Res. **18**(1), 39–50 (1981)
45. Hair, J.F., Jr., Howard, M.C., Nitzl, C.: Assessing measurement model quality in PLS-SEM using confirmatory composite analysis. J. Bus. Res. **109**, 101–110 (2020)
46. Cheung, G.W., Cooper-Thomas, H.D., Lau, R.S., Wang, L.C.: Reporting reliability, convergent and discriminant validity with structural equation modeling: a review and best-practice recommendations. Asia Pacific J. Manage. **41**(2), 745–783 (2024)
47. Kock, N.: Common method bias in PLS-SEM: a full collinearity assessment approach. Int. J. e-Collaboration (IJEC) **11**(4), 1–10 (2015)
48. Kock, N.: Measurement residual theory, analytic composites and the assessment of formative latent variables. J. Syst. Inf. Technol. (2025)
49. Acquaah, M.: Construct measurement in strategic management research in Africa. In: Advancing Research Methodology in the African Context: Techniques, Methods, and Designs, pp. 1–20. Emerald Group Publishing Limited (2014)
50. Bercovitz, J., Mitchell, W.: When is more better? The impact of business scale and scope on long-term business survival, while controlling for profitability. Strateg. Manag. J. **28**(1), 61–79 (2007)
51. Houston, M.B.: Assessing the validity of secondary data proxies for marketing constructs. J. Bus. Res. **57**(2), 154–161 (2004)
52. Spearman, C.: The measurement of intelligence. Eugenics Rev. **6**, 312–313 (1914)
53. Sarrigiannidis, S.O., Rey, J.M., Dobre, O., González-García, C., Dalby, M.J., Salmeron-Sanchez, M.: A tough act to follow: collagen hydrogel modifications to improve mechanical and growth factor loading capabilities. Mater. Today Bio **10**, 100098 (2021)
54. Sass, D.A.: Factor loading estimation error and stability using exploratory factor analysis. Educ. Psychol. Measur. **70**(4), 557–577 (2010)
55. Hair, J., Alamer, A.: Partial least squares structural equation modeling (PLS-SEM) in second language and education research: Guidelines using an applied example. Res. Methods Appl. Linguistics **1**(3), 100027 (2022)
56. Hair Jr, J.F., Hult, G.T.M., Ringle, C.M., Sarstedt, M., Danks, N.P., Ray, S., Ray, S., et al.: Evaluation of reflective measurement models. In: Partial Least Squares Structural Equation Modeling (PLS-SEM) Using R: A Workbook, pp. 75–90 (2021)
57. Hair Jr, J.F., Hult, G.T.M., Ringle, C.M., Sarstedt, M., Danks, N.P., Ray, S., Ray, S., et al.: Mediation analysis. Partial least Squares Structural Equation Modeling (PLS-SEM) Using R: A Workbook, pp. 139–153 (2021)
58. Howard, M.C., Henderson, J.: A review of exploratory factor analysis in tourism and hospitality research: identifying current practices and avenues for improvement. J. Bus. Res. **154**, 113328 (2023)
59. Valentini, F., Damasio, B.F.: Average variance extracted and composite reliability: reliability coefficients. Psicologia: Teoria e Pesquisa, **32**, e322225 (2016)
60. Alnoor, A., Atiyah, A.G., Abbas, S.: Blockchain adoption strategy from the perspective of organization, technology and environment theory: a hybrid SEM and ANN approach. Global Bus. Rev. 09721509241255482 (2024)
61. Henseler, J., Ringle, C.M., Sarstedt, M.: A new criterion for assessing discriminant validity in variance-based structural equation modeling. J. Acad. Mark. Sci. **43**, 115–135 (2015)
62. Khaw, K.W., Alnoor, A., Al-Abrrow, H., Chew, X., Sadaa, A.M., Abbas, S., Khattak, Z.Z.: Modelling and evaluating trust in mobile commerce: a hybrid three stage Fuzzy Delphi, structural equation modeling, and neural network approach. Int. J. Human-Comput. Interact. **38**(16), 1529–1545 (2022)
63. Alnoor, A.M., Al-Abrrow, H., Abdullah, H., Abbas, S.: The impact of self-efficacy on employees' ability to accept new technology in an Iraqi university. Glob. Bus. Organ. Excell. **39**(2), 41–50 (2020)

64. Aguirre-Urreta, M.I., Marakas, G.M., Ellis, M.E.: Measurement of composite reliability in research using partial least squares: some issues and an alternative approach. ACM SIGMIS Database: DATABASE Adv. Inf. Syst. **44**(4), 11–43 (2013)
65. Padilla, M.A., Divers, J.: A comparison of composite reliability estimators: coefficient omega confidence intervals in the current literature. Educ. Psychol. Measur. **76**(3), 436–453 (2016)
66. Cronbach, L.J.: Coefficient alpha and the internal structure of tests. Psychometrika, **16**(3), 297–334 (1951)
67. Carmines, E.G., Zeller, R.A.: Reliability and Validity Assessment. Sage publications (1979)
68. Nunnally, J.C.: An overview of psychological measurement. In: Clinical Diagnosis of Mental Disorders: A Handbook, pp. 97–146 (1978)
69. Lance, C.E., Butts, M.M., Michels, L.C.: The sources of four commonly reported cutoff criteria: What did they really say? Organ. Res. Methods **9**(2), 202–220 (2006)
70. Anderson, J.C., Gerbing, D.W.: Structural equation modeling in practice: a review and recommended two-step approach. Psychol. Bull. **103**(3), 411 (1988)
71. Albahri, A.S., Alnoor, A., Zaidan, A.A., Albahri, O.S., Hameed, H., Zaidan, B.B., Yass, A.A., et al.: Based on the multi-assessment model: towards a new context of combining the artificial neural network and structural equation modelling: a review. Chaos, Solitons Fractals **153**, 111445 (2021)
72. Li, Y., Wen, Z., Hau, K.T., Yuan, K.H., Peng, Y.: Effects of cross-loadings on determining the number of factors to retain. Struct. Equ. Modeling **27**(6), 841–863 (2020)
73. Dawson, J.F.: Moderation in management research: What, why, when, and how. J. Bus. Psychol. **29**(1), 1–19 (2014)
74. Locke, E.A., Shaw, K.N., Saari, L.M., Latham, G.P.: Goal setting and task performance: 1969–1980. Psychol. Bull. **90**(1), 125 (1981)
75. Vij, S., Farooq, R.: Moderating variables in business research. NLDIMSR Innovision J. Manage. Res. 31–44 (2017)
76. Arnold, H.J.: Moderator variables: a clarification of conceptual, analytic, and psychometric issues. Organ. Behav. Hum. Perform. **29**(2), 143–174 (1982)
77. Rouder, J.N., Engelhardt, C.R., McCabe, S., Morey, R.D.: Model comparison in ANOVA. Psychon. Bull. Rev. **23**, 1779–1786 (2016)
78. Hayes, A.F.: Partial, conditional, and moderated moderated mediation: quantification, inference, and interpretation. Commun. Monogr. **85**(1), 4–40 (2018)
79. Jamieson, J.: Analysis of covariance (ANCOVA) with difference scores. Int. J. Psychophysiol. **52**(3), 277–283 (2004)
80. Fiedler, K., Schott, M., Meiser, T.: What mediation analysis can (not) do. J. Exp. Soc. Psychol. **47**(6), 1231–1236 (2011)
81. Judd, C.M., Kenny, D.A.: Process analysis: estimating mediation in treatment evaluations. Eval. Rev. **5**(5), 602–619 (1981)
82. Bullock, J.G., Green, D.P., Ha, S.E.: Yes, but what's the mechanism? (don't expect an easy answer). J. Pers. Soc. Psychol. **98**(4), 550 (2010)
83. Zhao, X., Lynch, J.G., Jr., Chen, Q.: Reconsidering Baron and Kenny: myths and truths about mediation analysis. J. Consum. Res. **37**(2), 197–206 (2010)
84. Kenny, D.A., Baron, R.M.: The moderator-mediator variable distinction in social psychological research: conceptual, strategic and statistical considerations. J. Pers. Soc. Psychol. **51**(1), 1173–1182 (1986)

Part II
PLS-SEM and ANN

Chapter 3
Overview of ANN Analysis

Abstract This chapter aims to provide a simplified and comprehensive explanation of the SEM-PLS methodology. Data were collected from the Web of Science (Clarivate) base, which is a most reliable file. These studies were then refined based on their relevance to the presentation. A survey and critical analysis of these studies was conducted to identify the importance of the SEM-PLS methodology in arriving at the most accurate and authentic results.

Learning Objectives

1. This chapter aims to collect data in a survey manner and present it based on the nature of industry.
2. The chapter aims to provide a comprehensive overview of the literature on SEM-PLS in data analysis.
3. The chapter aims to provide a realistic understanding of the importance of the SEM-PLS approach in various industrial and service sectors.
4. This chapter enables the reader to understand the importance of the SEM-PLS approach in exploring linear and nonlinear relationships.
5. The chapter aims to provide a simplified model of how to employ approved statistical tools.

3.1 Introduction

For numerous periods, societal discipline researchers have been relying on numerical mechanisms and techniques for data examination drives. Advancements in computer technologies have significantly expanded the applications of statistical methods. Researchers initially relied on univariate and additional factor examines to comprehend information and their relations [1]. But, for a broader and in-depth understanding of the relationship[s] between variables, there is a need for advanced statistical methods. Furthermore, multivariate investigation includes the request of a

X. Chew et al., *Partial Least Squares Structural Equation Modeling and Complementary Methods in Business Research*, Information Systems Engineering and Management 67, https://doi.org/10.1007/978-3-032-01055-1_3

group of statistical approaches for simultaneously analyzing factors. So, a multivariate data examination contains a cluster of types signified by primary- and next-generation methods. Primary-generation approach includes regression-based methods, for example, linear regression, logistic regression, and variance investigation and investigative and assenting methods regarding measurement dependability [2]. Assuming the faintness of these methods, societal researcher favors next-generation methods, as structural equation modelling (SEM) [3]. SEM techniques facilitate researchers to incorporate those factors that are indirectly noticeable with indicator factors. SEM is categorized to two kinds. The primary one is called covariance-based SEM and is concerned with the refusal or receipt of theories after slightly group of methodical relations between factors which have been proven by experimentation. The next sort is called partial least squares (PLS) SEM or PLS track modelling, which provides theoretical explanations of the alteration of dependent factors when investigative models. SEM has developed a widely recognized and used technique in social research. Scientists crossways the sphere have made important projects by presenting numerous applications of much knowledge which are founded on next-generation methods [4]. Temporarily, a novel generation which combines SEM with Artificial Neural Networks (ANNs) has arose, and it is aimed at addressing the subjects that cannot be resolved by SEM method [5, 6]. With ANNs, the non-compensatory and nonlinear relations between factors can be revealed. Furthermore, the mixing between SEM and ANNs authenticates the validity of the results under SEM and encapsulates the non-linear relationships between backgrounds and result factors [7]. Hence, the ANN approach provides the vigorous validation of the important forecasters of a model [8]. To raise the rationality of forecasts for ANNs, existing studies have used influence approaches in deciding the importance of factors [4].

As a result, there has been a significant increase in studies that rely on SEM-ANN techniques for data investigation and analysis. However, a variety of cases involving specially made of different procedures in ANNs and influence approaches combination with SEM are unmoving unidentified, and this feature may affect the generation co nsequences. Therefore, more study is required. Hence, as a part of our ongoing research, there are some potential questions related to the disagreement over the use of SEM-ANN method, potential issues in using SEM-ANN and possible solutions etc. Scholars must address these questions to further encourage the use of SEM-ANN method.

The procedures connected to ANNs, and role approaches should be accessible based on two main orders, i.e., theoretical and practical. For the theory-based order, the studies in the literature should present dissimilar issues that effort to deliver the required answers to most of the glitches by decisive the import and relations of factors by means of the ANNs and role approaches. In these settings, the practical-based order should be coordinated and complemented by the theoretical-based order by decisive the forecast factors based on ANN, and the contribution processes should be discussed. Further, most current research attempted to use exact events and locations for the ANNs by using ten-fold, i.e. assortment of specific counts of the unseen layers and finding out kinds of initiation functions [9]. Hence, many construction assortment procedures are used for ANNs and contribution approaches, and they make sum-of-squares mistakes and varying fraction forecasts, assuming their dissimilar standards [10]. Therefore, the construction assortment procedures for ANN trials should be perfect to safeguard decisive consequences. Several studies assumed the compassion study to analyze the helps of factors and thus upsurge forecast correctness [11]. Previous studies have shown that ANN processes, through contribution calculation [12], are prone to pose other concerns and gaps in many aspects of their application.

Previous studies have not carefully addressed two exact challenges, i.e., (i) those related to the practical cases of ANNs (ii) those connected to influence approaches [13]. Hence, the several procedures of ANNs and contribution approaches for the identification of the most protuberant tests in the presence of unlike examination mechanisms also need to be studied in complexity. This debate has been augmented to a major issue about the efficiency of the technique, which is the impact of a piece procedure on the other, particularly as the literature has tried to address the linear and non-compensatory relations through dissimilar customisations of ANN and contribution processes [14, 15]. In this context, we raise another question i.e., what are the optional solutions for such stimulating features to choice the finest ANNs and contribution approaches in mixture with SEM?

A novel assortment procedure should be industrialized to overcome numerous problems in the customisation processes in ANNs and role approaches concerning their efficiency. Therefore, we increase the last important query of major contributions and objectives of the study under consideration? The main objectives of this study are to contribute to two objectives: (1) to current a comprehensive review of earlier studies on the practice of artificial neural network (ANNs) construction and structural equation modelling (SEM) and (2) to present a sequential procedure for doing ANN analysis.

3.2 Literature Survey

This study has been carried out under the defined criteria in the Preferred Reporting Items for Systematic Reviews and Meta-Analyses (PRISMA) (Fig. 3.1). PRISMA method relies on various databases for locating the relevant literature rather than relying on one database. Using a number of databases is appreciated because a single source does not contain enough research work to rely on. Hence, the usage of numerous databases ensures a wider coverage of relevant literature to demonstrate the theoretical understanding of the research issue.

To serve the purpose of our research, we relied on the four widely recognised literature sources, which are Scopus, Science Direct, The IEEE Xplore, and Web of Science (WoS). Our choice is because these four databases contain a large collection of studies conducted in a variety of fields, including physical and social sciences, science and technology and arts and humanities. These databases sufficiently served the purpose of our study.

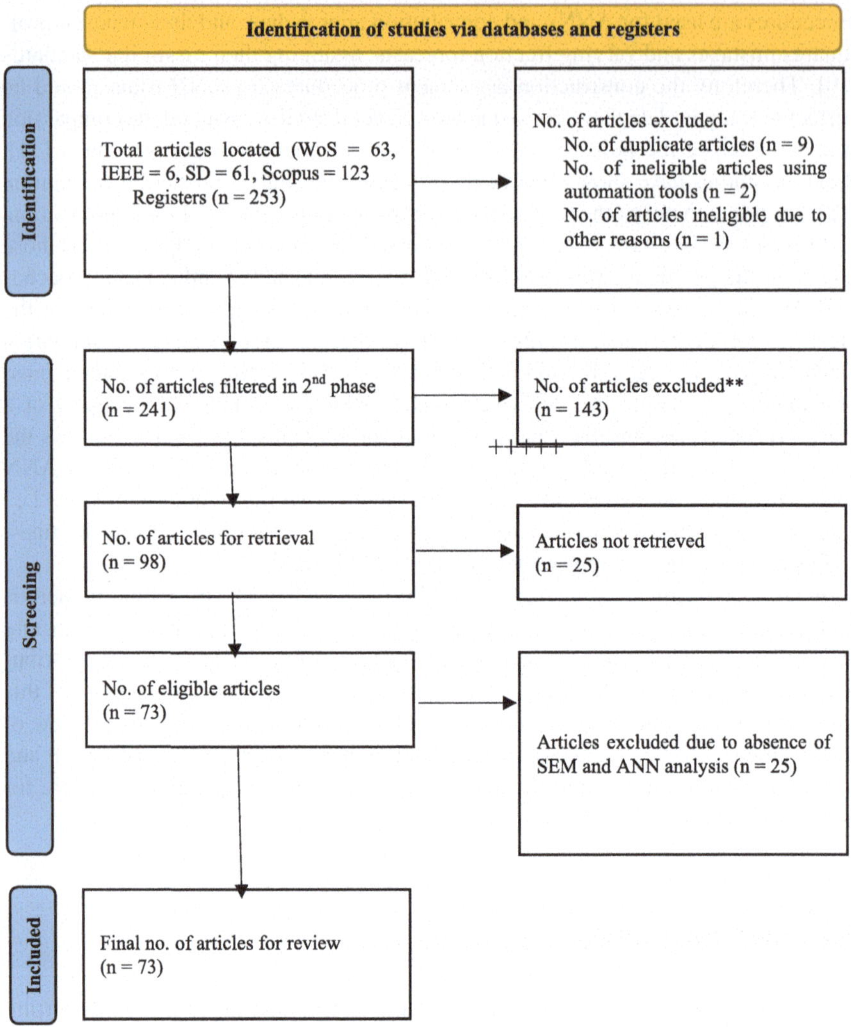

Fig. 3.1 Review protocol

For the selection of relevant studies, a comprehensive variety of articles was located from the above-stated four information sources. Once this was done, we excluded the identical and less pertinent research articles by interpreting names and abstracts. These selected research articles were thoroughly read with the purpose of choosing the most relevant articles. In this selection process, all six authors were involved in the first phase of locating the relevant studies, and in the second phase, five authors did the full text reading.

We used different prominent keywords to locate relevant literature in the above-mentioned databases. These prominent keywords included, for example, 'modelling, artificial neural networks, metadata, structural equation modelling, and so on. We limited the choice of articles only to the recent studies published in well-known research journals and renowned conference proceedings. This was done on the assumption that these two sources provided us with the most recent and latest research in the respective fields. A pre-defined criterion was used to make selection of the relevant research articles. Initially, we divided the relevant studies (SEM-ANN) into two categories i.e., general and coarse-grained classification. These two groups were industrialised from a survey of the works in the first stage of the articles collection. A three-point standard (i.e., journal/conference publication, it used SEM-ANN technique, it was published in English) was used to make selection of relevant articles.

From our initial search, we located 253 articles published between 2016 till April 2022 across all the four databases (IEEE Xplore: 6, ScienceDirect: 61, Scopus: 123, WoS: 63). After removing the articles beyond the research scope, the number of articles retained was reduced to 73, rendering to our presence standards. This project analysed 73 research studies from theoretical and practical perspectives; whilst classification consequences were got (Fig. 3.2).

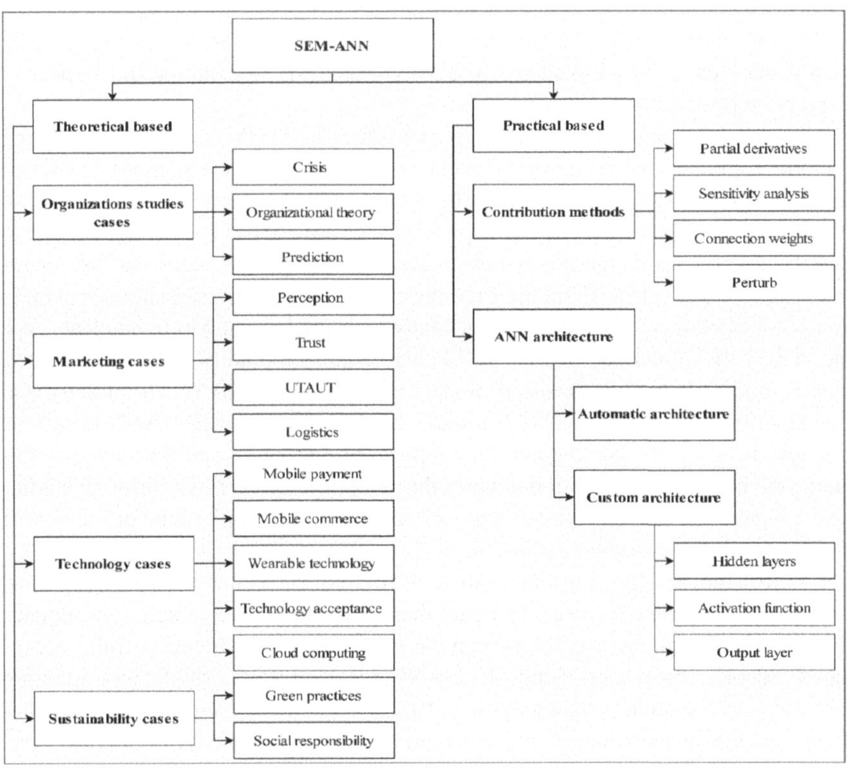

Fig. 3.2 Reviewed literature of SEM-ANN

As regards 'theoretically based' case studies, the 73 articles were secret giving to their isssues and comprise four subgroups, namely, organization of study, advertising, skill, and sustainability cases, thus determining the ratios of frequency deliveries among cases.

To 'practice-based, the 73 research studies were classified based on the ANNs and contribution approaches in mixture with SEM to realise the deep examination of the dissimilar customisation procedures.

3.3 ANN-SEM Previous Studies

The disaster works emphases on societal media, with numerous nations trying to curtail the terrible feast of sure worms. The articles show an energetic part in extra management, and they classify the issues swaying the intentions of societal media usage in crises by synthesising SEM and neural network examination [16]. Also, we discuss how organisational policies have emerged as a real problem for organisations because they create work weight, which touches workers' well-being. SEM–ANN has been used to control the issues that mostly affect removal from effort [11]. More deeply, another study adopted the SEM-ANN analysis to examine the impact of mergers on post-change strategies [17].

Moreover, the belongings of background issues on (TQM) and (HR), assuming that little excellence is an additional actual issues met by societies among the disappointment to doing (ISO) packages. In deciding the background aspects and an instrument for implementing (TQM) packages, (SEM–ANN) has been practiced to classify the greatest dangerous issues to switch structural efficiency and its connotation with (TQM) [18]. Then, the examples in which numerous businesses expression glitches connected to forecasting. Many vicissitudes can barely be clear, even uncertainty they touch the overall well-being of the community. To deliver expectedness, studies have industrialised models based on (SEM–ANN) to classify and look at unforeseen cases [19–22]. Similarly, Shukla [23] used SEM-ANN technique to study the students' motivations to adopt mobile learning and the factors influencing the behavioural intention towards this adoption. Conversely, different studies have adopted the same analytical approach SEM–ANN to explore the possibility of applying digital technologies [24–26].

Based on the literature, customers are found to frequently change their perceptions. Usually, it is not easy for firms to retain their customers. This situation designates that client expectations and the perception of making way should be fully appreciated. Scholars have been using SEM–ANN for determining the factors affecting customers' expectations and perceptions [10, 12, 27, 28]. In the same context, another study used this analytical approach to examine the effect of (eWOM) on customers' choice [24, 25, 29, 30]. Trust is a source of great concern, especially in the case of online shopping, because there are numerous factors which may influence online purchase intentions. Several factors may affect the customers' trust in e-commerce. For the identification of these factors and to address the potential issues. The use of (SEM–ANN) is highly recommended to unveil the factors which are pertinent to customers' belief in e-trade and as an income to provision e-shop forums [31–34]. The (UTAUT) model argues the issues that influence new technology adoption. SEM–ANN technique founded on (UTAUT2) has been rummage-sale to forecast the technological acceptance issues by clienteles [9, 18, 35–41]. Regarding logistics, focus on the supplier chain as a strategical tool to attain action. The (SEM–ANN) technique has been rummage-sale not lone to highlight important supplier chain does but also to recover firm act [42, 42]. SEM–ANN has been adopted in mobile payment to describe the chief factors stopping clients from utilizing electronic payments and the factors swaying their choice of technology adoption [32, 43–48]. Mobile commerce is are issue for many firms concerning the absence of faith towards clienteles because of the many efforts to enter the market. Scholars adopted a two-step technique of SEM-ANN to explain the decision-making process of smartphone buyback. Similarly, the SEM-ANN technique has been adopted to highlight the major factors influencing consumers' intentions to adopt mobile commerce. Literature has also cited studies about the effect of social media advertisement on customers' online purchase intentions [4, 49–57]. Regarding clothing technology, the research cites that clothing healthcare technology has aided older countries in retrieving healthcare information. However, a problem exists in terms of resistance to the adoption of wearable healthcare technology by elderly people. To address such tasks, the (SEM) factors may be rummage to identify the essential factors based on ANN and thereby facilitate elderly people towards technology adoption [58–63]. Technology acceptance signifies the problematic of technological acceptance of many organisations because of an absence of info of the issues that touch individuals in adopting new skills. (SEM–ANN) has been used to highpoint these issues and join to the problem of technology adoption [5, 64–68]. In medical studies, TMA has been accepted to discover the causes of smarttools use in healthcare. Many organisations attempt to accept cloud computing. Directors are finally astonished by a resistance to change. SEM–ANN has been used to travel the main pointers stopping the acceptance of technics in organisations [18, 69, 70].

ANN and theories.

3.4 ANN and TOE Framework and Inspiration Theory

Although many theories have been validated using ANNs, these two are considered among the most important and influential, especially since TOE governs the acceptance and adoption of new technology. Inspiration theory, which consists of personality and content characteristics, has occupied a significant place in social commerce research and is considered a modern theory [71]. Both of these theories provide solid theoretical depth in the process of analyzing technology acceptance. They also help clarify the impact of technology on changing consumer behavior, especially in the digital age. [72] are theorists of the TOE concept, which consists of three components: organization, technology, and environment. They believe that these three components can collectively influence the decision to adopt technology in organizations, particularly in education and healthcare institutions. In this context, Alnoor et al. [19] studied the TOE framework to understand the applicability of blockchain. They analyzed environmental, technological, and organizational components by integrating ANNs. This approach allows for the identification of nonlinear relationships, i.e., indirect causation, in addition to the direct causation identified by PLS-SEM. The inclusion of the three dimensions in the TOE framework makes it comprehensive [73]. This allows researchers to study these elements with a deeper insight into the institutional reality of adopting new technology. ANN technology contributes to drawing clearer impact maps, enabling researchers to understand causal factors as well as predictive aspects. Traditional analytical techniques such as regression models cannot achieve these results. Therefore, ANN technology is an effective element in obtaining the most accurate results. Inspiration theory emerged in [74] through the insights of Böttger et al. Inspiration theory is an effective element in the fields of social commerce and consumer behavior. The authors of this theory see it as a definition of how inspiration is employed in commerce and its role in transforming consumer behavior from one state to another. Inspiration is an external motivator that is triggered to achieve a specific goal. In the field of commerce, consumer behavior is the target of certain motivational elements. Attractiveness, credibility, and trust are all included under the term extrinsic motivators. In this context, a study [63] revealed that credibility plays a role in attracting customers to purchase from digital platforms. This type of behavior is sometimes called "positive content-driven behaviour" [75]. Because credibility generates trust in customers, they are driven to write positive feedback as they continue to interact. It can be argued that these two theories have expanded research on e-commerce and consumer behavior, facilitating strategic decisions about technology adoption. They have also explained consumer behavioral shifts.

References

1. Hair, J.F., Jr., Sarstedt, M., Hopkins, L., Kuppelwieser, V.G.: Partial least squares structural equation modeling (PLS-SEM): an emerging tool in business research. Eur. Bus. Rev. **26**(2), 106–121 (2014)
2. Smith, D., Hair, J.F., Jr., Ferguson, K.: An investigation of the effect of family influence on commitment-trust in retailer–vendor strategic partnerships. J. Fam. Bus. Strat. **5**(3), 252–263 (2014)
3. Alnoor, A.: Human capital dimensions and firm performance, mediating role of knowledge management. Int. J. Bus. Excell. **20**(2), 149–168 (2020)
4. Khaw, K.W., Alnoor, A., Al-Abrrow, H., Chew, X., Sadaa, A.M., Abbas, S., Khattak, Z.Z.: Modelling and evaluating trust in mobile commerce: a hybrid three stage fuzzy Delphi, structural equation modeling, and neural network approach. Int. J. Human-Comput. Interact. **38**(16), 1529–1545 (2022)
5. Ahani, A., Rahim, N.Z.A., Nilashi, M.: Forecasting social CRM adoption in SMEs: a combined SEM-neural network method. Comput. Hum. Behav. **75**, 560–578 (2017)
6. Asadi, S., Nilashi, M., Samad, S., Rupani, P.F., Kamyab, H., Abdullah, R.: A proposed adoption model for green IT in manufacturing industries. J. Clean. Prod. **297**, 126629 (2021)

7. Alnoor, A., Atiyah, A.G.: An SEM-ANN analysis of the influence of strategic merger on strategic postchange of family businesses. Nankai Bus. Rev. Int. (ahead-of-print) (2024)
8. Al-Abrrow, H., Alnoor, A., Abbas, S.: The effect of organizational resilience and CEO's narcissism on project success: organizational risk as mediating variable. Organ. Manage. J. **16**(1), 1–13 (2019)
9. Chen, C.C., Tsang, S.S.: Predicting adoption of mobile payments from the perspective of taxi drivers. IET Intel. Transport Syst. **13**(7), 1116–1124 (2019)
10. Sharma, A., Dwivedi, Y.K., Arya, V., Siddiqui, M.Q.: Does SMS advertising still have relevance to increase consumer purchase intention? A hybrid PLS-SEM-neural network modelling approach. Comput. Hum. Behav. **124**, 106919 (2021)
11. Abubakar, A. M., Ilkan, M., Al-Tal, R. M., Eluwole, K. K.: eWOM, revisit intention, destination trust and gender. J. Hospital. Tour. Manage. **31**, 220–227 (2017)
12. Hew, J.J., Leong, L.Y., Tan, G.W.H., Lee, V.H., Ooi, K.B.: Mobile social tourism shopping: a dual-stage analysis of a multi-mediation model. Tour. Manage. **66**, 121–139 (2018)
13. Hsu, S.H., Chen, W.H., Hsieh, M.J.: Robustness testing of PLS, LISREL, EQS and ANN-based SEM for measuring customer satisfaction. Total Qual. Manag. Bus. Excell. **17**(3), 355–372 (2006)
14. Najmi, A., Kanapathy, K., Aziz, A.A.: Exploring consumer participation in environment management: findings from two-staged structural equation modelling-artificial neural network approach. Corp. Soc. Responsib. Environ. Manag. **28**(1), 184–195 (2021)
15. Najmi, A., Kanapathy, K., Aziz, A.A.: Understanding consumer participation in managing ICT waste: findings from two-staged structural equation modeling-artificial neural network approach. Environ. Sci. Pollut. Res. **28**, 14782–14796 (2021)
16. Talwar, M., Talwar, S., Kaur, P., Islam, A. N., Dhir, A.: Positive and negative word of mouth (WOM) are not necessarily opposites: A reappraisal using the dual factor theory. J. Retail. Cons. Ser. **63**, 102396 (2021)
17. Alnoor, A., Atiyah, A.G.: An SEM-ANN analysis of the influence of strategic merger on strategic postchange of family businesses. Nankai Bus. Rev. Int. (ahead-of-print) (2025)
18. Alam, M.Z., Hu, W., Kaium, M.A., Hoque, M.R., Alam, M.M.D.: Understanding the determinants of mHealth apps adoption in Bangladesh: a SEM-Neural network approach. Technol. Soc. **61**, 101255 (2020)
19. Alnoor, A., Abbas, S., Sadaa, A.M., Chew, X., Bayram, G.E.: Navigating the power of blockchain strategy: analysis of technology-organization-environment (TOE) framework and innovation resistance theory using PLS-SEM and ANN insights. Technol. Forecast. Soc. Chang. **214**, 124044 (2025)
20. Prasad, S., Singh, A., Korres, N. E., Rathore, D., Sevda, S., Pant, D.: Sustainable utilization of crop residues for energy generation: A life cycle assessment (LCA) perspective. Biores. Technol. **303**, 122964 (2020)
21. Kunnapapdeelert, S., Pitchayadejanant, K.: Hybrid SEM-neural networks for predicting electronics logistics information system adoption in Thailand healthcare supply chain. Int. J. Buss. Perform. Supply Chain Model. **11**(1), 54–68 (2020)
22. Kruk, M., Paturej, E., Artiemjew, P.: From explanatory to predictive network modeling of relationships among ecological indicators in the shallow temperate lagoon. Ecol. Indic. **117**, 106637 (2020)
23. Shukla, R., Kumar, P., Vishwakarma, D. K., Ali, R., Kumar, R., Kuriqi, A.: Modeling of stage-discharge using back propagation ANN-, ANFIS-, and WANN-based computing techniques. Theo. Appl. Climatol. **147**(3), 867–889 (2022)
24. Alnoor, A., Atiyah, A.G., Abbas, S.: Unveiling the determinants of digital strategy from the perspective of entrepreneurial orientation theory: a two-stage SEM-ANN approach. Glob. J. Flex. Syst. Manag. **25**(2), 243–260 (2024)
25. Alnoor, A., Tiberius, V., Atiyah, A.G., Khaw, K.W., Yin, T.S., Chew, X., Abbas, S.: How positive and negative electronic word of mouth (eWOM) affects customers' intention to use social commerce? A dual-stage multi group-SEM and ANN analysis. Int. J. Human-Comput. Interact. **40**(3), 808–837 (2024)

26. XinYing, C., Tiberius, V., Alnoor, A., Camilleri, M., Khaw, K.W.: The dark side of metaverse: a multi-perspective of deviant behaviors from PLS-SEM and fsQCA findings. Int. J. Human-Comput. Interact. **41**(5), 3128–3148 (2025)
27. Song, M., Qiao, L., Law, R.: Formation path of customer engagement in virtual brand community based on back propagation neural network algorithm. Int. J. Comput. Sci. Eng. **22**(4), 454–465 (2020)
28. Hew, J. J., Leong, L. Y., Tan, G. W. H., Ooi, K. B., Lee, V. H.: The age of mobile social commerce: An artificial neural network analysis on its resistances. Technol. Forecast. Soc. Change **144**, 311–324 (2019)
29. Cai, S., Liu, Y., Aduldecha, S., Junaidi, J.: Leveraging customer green behavior toward green marketing mix and electronic word-of-mouth. Sustainability **17**(6), 2360 (2025)
30. Wanigapura, T.M., Guruge, T.P.S.R., Kuruppu, I.V., Abeysiriwardana, P.C.: Diversified impact of electronic word-of-mouth (eWOM) on consumer communities: a developing country perspective. DECISION, 1–17 (2025)
31. Kalinić, Z., Marinković, V., Kalinić, L., Liébana-Cabanillas, F.: Neural network modeling of consumer satisfaction in mobile commerce: an empirical analysis. Expert Syst. Appl. **175**, 114803 (2021)
32. Leong, L.Y., Hew, T.S., Ooi, K.B., Chong, A.Y.L.: Predicting the antecedents of trust in social commerce—a hybrid structural equation modeling with neural network approach. J. Bus. Res. **110**(1), 24–40 (2020)
33. Sayem, S.M., Islam, A., Uddin, M.R., Promy, J.S.: Determinants of e-commerce customer satisfaction: mediating role of IT innovation acceptance. Int. J. Qual. Reliab. Manage. **42**(1), 86–106 (2025)
34. Yu, T., Teoh, A.P., Bian, Q., Liao, J., Wang, C.: Can virtual influencers affect purchase intentions in tourism and hospitality e-commerce live streaming? An empirical study in China. Int. J. Contemp. Hosp. Manag. **37**(1), 216–238 (2025)
35. Abbasi, G.A., Tiew, L.Y., Tang, J., Goh, Y.N., Thurasamy, R.: The adoption of cryptocurrency as a disruptive force: deep learning-based dual stage structural equation modelling and artificial neural network analysis. PLoS ONE **16**(3), e0247582 (2021)
36. Alkawsi, G.A., Ali, N., Mustafa, A.S., Baashar, Y., Alhussian, H., Alkahtani, A., Tiong, S.K., Ekanayake, J.: A hybrid SEM-neural network method for identifying acceptance factors of the smart meters in Malaysia: challenges perspective. Alex. Eng. J. **60**(1), 227–240 (2021)
37. Duan, S.X., Deng, H.: Hybrid analysis for understanding contact tracing apps adoption. Ind. Manag. Data Syst. **121**(7), 1599–1616 (2021)
38. Sharma, S.K., Al-Badi, A., Rana, N.P., Al-Azizi, L.: Mobile applications in government services (mG-App) from user's perspectives: a predictive modelling approach. Gov. Inf. Q. **35**(4), 557–568 (2018)
39. Wang, G., Tan, G.W.H., Yuan, Y., Ooi, K.B., Dwivedi, Y.K.: Revisiting TAM2 in behavioral targeting advertising: a deep learning-based dual-stage SEM-ANN analysis. Technol. Forecast. Soc. Chang. **175**, 121345 (2022)
40. Wu, H., Que, X., Pan, L.: Analysing students' aspiration factors that impact actual use of mobile learning systems: a two-stage SEM-ANN approach. Humanit. Soc. Sci. Commun. **12**(1), 1–15 (2025)
41. Xhafaj, E., Qendraj, D.H., Kosova, R., Gjikaj, N., Mersinllari, O., Alikaj, L.S.: An integrating framework through the extension of the UTAUT2 model for online banking: a context from a two-staged approach with PLS-SEM and Fuzzy Z-AHP. Eng. Econ. **36**(2), 241–254 (2025)
42. Soomro, R.B., Al-Rahmi, W.M., Dahri, N.A., Almuqren, L., Al-Mogren, A.S., Aldaijy, A.: A SEM–ANN analysis to examine impact of artificial intelligence technologies on sustainable performance of SMEs. Sci. Rep. **15**(1), 5438 (2025)
43. Liébana-Cabanillas, F., Marinkovic, V., De Luna, I.R., Kalinic, Z.: Predicting the determinants of mobile payment acceptance: a hybrid SEM-neural network approach. Technol. Forecast. Soc. Chang. **129**, 117–130 (2018)
44. Loh, X.M., Lee, V.H., Hew, T.S., Lin, B.: The cognitive-affective nexus on mobile payment continuance intention during the COVID-19 pandemic. Int. J. Bank Market. **40**(5), 939–959 (2022)

45. Ooi, K.B., Tan, G.W.H.: Mobile technology acceptance model: an investigation using mobile users to explore smartphone credit card. Expert Syst. Appl. **59**, 33–46 (2016)
46. Walle, Y.M.: Deep learning-based hybrid SEM-neural network approach for predicting mobile application adoption in whistleblowing services. Int. J. Disclosure and Governance 1–14 (2025)
47. Alam, J., Raza, M. A., Muzzammil, M.: Efficacy of artificial neural network and adaptive neuro-fuzzy inference system models in scour depth prediction at submerged weirs. J. Water Manage. Model. (2025)
48. Sharma, S. K., Sharma, M.: Examining the role of trust and quality dimensions in the actual usage of mobile banking services: An empirical investigation. Int. J. Inf. Manag. **44**, 65–75 (2019)
49. Ali Abbasi, G., Abdul Rahim, N.F., Wu, H., Iranmanesh, M., Keong, B.N.C.: Determinants of SME's social media marketing adoption: competitive industry as a moderator. SAGE Open **12**(1), 21582440211067220 (2022)
50. Geetha, V.: Students' behavioural intention towards mobile technology for online shopping. Int. J. Electron. Market. Retail. **16**(1), 67–78 (2025)
51. Khalek, S.A., Dey, D., Chakraborty, A., Samanta, T.: "From expectations to frustrations": dissecting negative experiences to understand negative word-of-mouth in online grocery services. J. Retail. Consum. Serv. **84**, 104221 (2025)
52. Lee, V.H., Hew, J.J., Leong, L.Y., Tan, G.W.H., Ooi, K.B.: Wearable payment: a deep learning-based dual-stage SEM-ANN analysis. Expert Syst. Appl. **157**, 113477 (2020)
53. Liébana-Cabanillas, F., Marinković, V., Kalinić, Z.: A SEM-neural network approach for predicting antecedents of m-commerce acceptance. Int. J. Inf. Manage. **37**(2), 14–24 (2017)
54. Liébana-Cabanillas, F., Singh, N., Kalinic, Z., Carvajal-Trujillo, E.: Examining the determinants of continuance intention to use and the moderating effect of the gender and age of users of NFC mobile payments: a multi-analytical approach. Inf. Technol. Manage. **22**(2), 133–161 (2021)
55. Nim, N., Jeong, Y., Martinez, J.F., Smith, L.: Retailing in metaverse: cryptocurrency and consumer payment choices in virtual reality environments. J. Retail. (2025)
56. Ray, A., Bala, P.K., Rana, N.P.: Exploring the drivers of customers' brand attitudes of online travel agency services: a text-mining based approach. J. Bus. Res. **128**, 391–404 (2021)
57. Sharma, S.K., Joshi, A., Sharma, H.: A multi-analytical approach to predict the Facebook usage in higher education. Comput. Hum. Behav. **55**, 340–353 (2016)
58. Alam, M.M.D., Alam, M.Z., Rahman, S.A., Taghizadeh, S.K.: Factors influencing mHealth adoption and its impact on mental well-being during COVID-19 pandemic: a SEM-ANN approach. J. Biomed. Inform. **116**, 103722 (2021)
59. Asadi, S., Abdullah, R., Safaei, M., Nazir, S.: An integrated SEM-neural network approach for predicting determinants of adoption of wearable healthcare devices. Mob. Inf. Syst. **2019**(1), 8026042 (2019)
60. Capulli, E., Druda, Y., Palmese, F., Butt, A.H., Domenicali, M., Macchiarelli, A.G., Ingravallo, F., et al.: Ethical and legal implications of health monitoring wearable devices: a scoping review. Soc. Sci. Med. 117685 (2025)
61. Chandrasekaran, R., Sadiq, T.M., Moustakas, E.: Usage trends and data sharing practices of healthcare wearable devices among US adults: cross-sectional study. J. Med. Internet Res. **27**, e63879 (2025)
62. Talukder, M.S., Sorwar, G., Bao, Y., Ahmed, J.U., Palash, M.A.S.: Predicting antecedents of wearable healthcare technology acceptance by elderly: a combined SEM-Neural Network approach. Technol. Forecast. Soc. Chang. **150**, 119793 (2020)
63. Yang, P., Bi, G., Qi, J., Wang, X., Yang, Y., Xu, L.: Multimodal wearable intelligence for dementia care in healthcare 4.0: a survey. Informat. Syst. Front. **27**(1), 197–214 (2025)
64. Almarzouqi, A., Aburayya, A., Salloum, S.A.: Determinants of intention to use medical smartwatch-based dual-stage SEM-ANN analysis. Inform. Med. Unlocked **28**, 100859 (2022)
65. Basit, A., Wang, L., Javed, A., Shoaib, M., Aslam, M.U.: Impact of digital technologies on manufacturing firm resilience during COVID-19 pandemic: a PLS-SEM and artificial neural network analysis. J. Manuf. Technol. Manag. **36**(2), 358–384 (2025)

66. Nguyen Thi Ha, P., Tran Van, H., Le Thai, P., Trinh Thi Phuong, T., Tran, T.: Partial least squares structural equation modeling (PLS-SEM) in higher education research: an evidence from using technology acceptance model (TAM) and innovation resistance theory (IRT). New Rev. Inf. Network. 1–31 (2025)
67. Songkram, N., Chootongchai, S.: Adoption model for a hybrid SEM-neural network approach to education as a service. Educ. Inf. Technol. **27**(5), 5857–5887 (2022)
68. Lee, K. Y., Chung, N., Hwang, S.: Application of an artificial neural network (ANN) model for predicting mosquito abundances in urban areas. Ecol. Info. **36**, 172–180 (2016)
69. Khayer, A., Talukder, M.S., Bao, Y., Hossain, M.N.: Cloud computing adoption and its impact on SMEs' performance for cloud supported operations: a dual-stage analytical approach. Technol. Soc. **60**, 101225 (2020)
70. Ooi, K.B., Lee, V.H., Tan, G.W.H., Hew, T.S., Hew, J.J.: Cloud computing in manufacturing: the next industrial revolution in Malaysia? Expert Syst. Appl. **93**, 376–394 (2018)
71. Song, S., Yao, X., Zhao, Y.C., Ba, Z.: Get inspired and pay for the goods: an investigation of customer inspiration and purchase intention in livestream shopping. J. Retail. Consum. Serv. **78**, 103750 (2024)
72. Tornatzky, L.G., Fleischer, M.: The Processes of Technological Innovation. Lexington Books (1990)
73. Dadhich, M., Hiran, K.K.: Empirical investigation of extended TOE model on corporate environment sustainability and dimensions of operating performance of SMEs: a high order PLS-ANN approach. J. Clean. Prod. **363**, 132309 (2022)
74. Böttger, T., Rudolph, T., Evanschitzky, H., Pfrang, T.: Customer inspiration: conceptualization, scale development, and validation. J. Mark. **81**(6), 116–131 (2017). https://doi.org/10.1509/jm.15.0006
75. Lambert, C., Choi, F., Chandrasekharan, E.: Positive reinforcement helps breed positive behavior": moderator perspectives on encouraging desirable behavior. Proc. ACM Human-Comput. Interact. **8**(CSCW2), 1–33 (2024)
76. Alshammari, S., Aichouni, M., Ben Ali, N., Alshammari, O., Alfaraj, F., Aichouni, A.B.E.: Impact of total quality management and lean manufacturing on sustainability performance: an SEM-ANN approach in saudi food manufacturing (2025)

Chapter 4
Business Research Applications in SEM-ANN Analysis

Abstract This chapter provides a detailed overview of the importance of the PLS-SEM methodology in scientific research. It reviews the role of this analytical methodology in studying the relationships between variables to arrive at the most accurate results. Furthermore, a section is devoted to comparing PLS-SEM with other analytical methodologies, such as CB-SEM, which are used in the same research contexts. The chapter also explains the importance of the methodology in examining previous studies. This examination is based on what the researcher wants to know. For example, classifying studies according to their chronological order or recency, their geographical location, and the number of times they are cited. Furthermore, the methodology is important in providing analyses supported by figures and graphs.

Learning Objectives

1. This chapter aims to provide a detailed overview of data screening before testing.
2. Providing a comprehensive explanation of the importance of the SEM method in scientific research.
3. Providing a comprehensive explanation of the importance of the SEM method in scientific research.
4. Identifying scientific production of (SEM) models using artificial neural networks (ANN) based on countries contributes.
5. Understanding the research taxonomy, which exemplifies and substantiates that the greatest research that joint the SEM-ANN.

4.1 Introduction

For more than a century, statistical analysis with its simple applications has been an important tool for studies, especially in economics and social sciences. Computers are the main engine and continuous supporter of the development of statistical methods. These technologies have enabled the development of very easy and fast applications.

X. Chew et al., *Partial Least Squares Structural Equation Modeling and Complementary Methods in Business Research*, Information Systems Engineering and Management 67, https://doi.org/10.1007/978-3-032-01055-1_4

At first, studies tended towards univariate analyses. For example, they are studying the level of poverty in some countries in North Africa [1]. Or other studies that are not concerned with the association of two variables with each other. However, with the development of technological capabilities and the expansion of economic and social issues, studies turned towards investigating the relationships between two variables. Until they reached what they are today in terms of the possibility of exploring more complex relationships. In addition to expanding the scope of the societies studied to reach thousands in many studies. This is thanks to the development of statistical techniques and their application programs. The process of analyzing multivariate studies continues to this day. At the same time, the statistical methods used are developing in line with the nature of the analysis of these variables. Analysis is always linked to the direction of the relationships between variables. This has led to the expansion of statistical models at the level of theory and application. Variables often tend to measure issues related to companies, individuals, activities, ... etc. Therefore, researchers work to collect data from reliable and irrefutable sources, by relying on questionnaires (primary data) or databases for the activity in question (secondary data).

Previous analysis methods used techniques called first-generation techniques [2]. These techniques are usually used in the study of social sciences, to express two variables, one of which is influential and the other is influenced. These techniques adopted two types of analysis, namely exploratory factor analysis and confirmatory factor analysis. The first is used to explore a theory that has not yet been studied. Therefore, it is used when there is no prior answer about the nature of the relationship between the variables to be studied. Among these techniques are analysis of variance and linear regression analysis. As for the second (confirmatory factor analysis), it is used to confirm the applicability of previously studied theories. By confirming the patterns of relationships between the variables arising from those theories. Among the applications used in this type of analysis are cluster analysis and multidimensional scaling. With the continued use of these two techniques in analysis, a controversy has emerged in the research community. This controversy can be summarized in the lack of clarity or lack of distinction between exploratory analysis and confirmatory analysis. Due to the overlap between the two concepts in one model on the one hand. Or due to the imbalance of influence between the variables on the other hand. For example, researchers use the regression analysis method to prove the influence of variables on each other. At the same time, the same analysis can be used to explore the possibility of expanding the model across the independent variables themselves. On the other hand, when exploring the relationships between variables for the purpose of determining their structural composition and unifying their elements. Their complex factors are gathered and unified on the basis of the relationships between the variables. Based on the studied data. Therefore, most researchers in the field of social sciences moved to second-generation techniques [3]. Especially at the beginning of the twenty-first century. To overcome the weaknesses found in primary-group methods.

The secondary-group methods are called structural equation modeling (SEM). SEM techniques are divided into two types. The first type is covariance-based SEM (CB-SEM), and the second type is partial least squares (PLS-SEM). The first type is used to accept or reject the theory being tested. In other words, it supports conducting experimental studies to estimate the relationships between a set of variables. Through the ability of the proposed model to estimate the covariance matrix for the studied sample. In contrast, the second type is used to strengthen theories based on exploratory studies. By explaining the variance in the dependent variable/variables. So, it can be said that SEM methods help researchers integrate elements that cannot be observed and measured by indicator components. It also enables the measurement error to be easily calculated [4].

4.2 Importance of SEM in Research

SEM offers several advantages that make it of great importance in analyzing research models. It is capable of analyzing complex relationships between multiple variables simultaneously. SEM technique is capable of examining both observable variables and latent variables. This allows for a broader and deeper understanding of the study variables and the model as a whole. SEM technique allows testing previously studied theories as a confirmatory approach. SEM can also assess the suitability of the data to the study model. SEM enables researchers to assess the validity and reliability of the study scale. Based on the hypothesized model which the relations among the latent variables and the observed variables appear. This helps in assessing measurement errors and assessing the reliability of the study scale. The SEM approach is characterized by the presence of several indicators that can be used to assess latent variables. Thus, reducing measurement errors. Therefore, we find SEM successful in complex studies that require more than one indicator in measurement. SEM provides the researcher with the ability to test complex models that have more than one path (independent, mediator, dependent, moderator) to test direct and indirect effects. The SEM approach integrates several analyses (factor analysis, path analysis, and regression analysis) into one integrated analysis. This type of analysis provides a comprehensive understanding of the model. SEM is flexible in application. It can be applied to longitudinal, cross-sectional, and multi-group studies. Hence, it is suitable for a variety of data. Finally, SEM is a powerful and comprehensive tool for examination and analysis in various social sciences.

4.3 PLS-SEM Versus CB-SEM

In the structural relationship model, there are two methods to estimate the level of relationships between variables. The first method is (CB-SEM) and the second is (PLS-SEM). PLS-SEM is the focus of our book. It is the most recent and widely used method at the present time. It is important for those interested to understand the difference between each method. In order to choose the best method for analysis. The main difference between the CB-SEM and PLS-SEM approaches is the way in which the latent variables are treated. CB-SEM considers the latent variables as constructs that explain the magnitude of the variance between the indicators associated with them. In contrast, PLS-SEM adopts weighted constructs of the indicator variables associated with a particular structure. They serve as proxies to represent the structures of interest. Therefore, the PLS-SEM approach is called the non-assumption approach as in CB-SEM. PLS-SEM reduces the strength of the assumptions of CB-SEM, which sees that the variance between the indicators is explained by a common factor [5]. The use of weighted composites for the indicator structures also helps in calculating measurement error. This makes the PLS-SEM approach more successful in estimating multiple regression coefficients [6]. Thus, if there is a model consisting of ten variables, all weights are set to 1. Therefore, the mathematical formula for the composite variables resulting from the linear combination will be as follows:

$$\text{Composite value} = 1 \cdot x1 + 1 \cdot x2 + 1 \cdot x3 + 1 \cdot x4 \ldots + 1 \cdot x10$$

For example, if the respondent obtained scores from 1 to 10 for the ten variables respectively, the corresponding composite value will be 10. This means that there is a clear ease in estimating and applying the regression coefficients. Which usually cannot be ignored as it is considered bias.

4.4 Key Consideration Before Data Analysis

For the researcher, to succeed in obtaining accurate and reliable results, must verify the data [7]. Descriptive analyses can be relied upon, which are usually available in SPSS version 23 or some other versions. These tools are used to scan and organize data, which helps the researcher to proceed with the work after checking the data well and deleting the numbers that are outside the range of probability [8]. The frequency distribution is used to examine the nature of the data and identify data that is outside the range [9]. For example, if you use the seven-point Likert scale and you code your data from 1 to 7, then any value greater or smaller than the mentioned values is considered outside the range of probability.

When the researcher distributes the study questionnaire and then collects it, there is a possibility that some of it will lose part of the information required to be filled out by the respondent, for several reasons, including not filling out all the questionnaire paragraphs correctly, or the possibility of an error in the process of entering the data for the analysis [10]. If this loss of some data occurs at significant rates, this will affect the credibility of the statistical analysis of it, in addition to the fact that some programs do not work when there is missing data. As a result, it is possible to use the statistics package (Spss, V.23) to explore the percentages of missing data, using the frequency method. Such a step is sufficient to determine the number of missing data from valid data. Hair et al. [10] argued that when the percentages of missing data exceed 15%, they distort the results of the study.

The researcher's selection of the appropriate statistical tools for his field of study is one of the most important requirements for obtaining accurate and expressive results. Within the statistical field, there are binary types of data distribution: the normal and the non-normal. The importance of determining the type of distribution leads us to select parametric or non-parametric statistics [11, 12]. So, verifying the nature of the data distribution is a necessary step that directs the researcher toward the appropriate statistics to analyze his data. When the data distribution appears to be normal, this means that the researcher must use parametric statistical tools. However, if the data appears to follow the non-normal distribution, this means that the researcher must use non-parametric statistical tools. It is noted that it is often preferable to use parametric statistics in hypothesis testing, and the reason is that these statistics include distinctive properties over non-parametric statistics, especially in the scope of statistical power and ease of implementation and estimation [11]. According to Hair et al. [13], the Kolmogorov–Smirnov and Shapiro-Wilks methods do not provide sufficient guidance about the nature of the data distribution, so "Kurtosis & Skewness" can be adopted because they provide an accurate analysis of the data derived from the study samples [14].

Kurtosis & Skewness method helps to know whether the answers to the questionnaire are within the normal curve or not, and the (Skewness) test indicates the deviation (extent of skewness) of the data from the natural center of the answers, and it is either a right deviation or a left deviation from the center of the answers, and the normal distribution is ideal at point (0), and it has a positive deviation when all the answers are to the left of the center, and the deviation is negative when all the answers are to the right of the main center of the answers. The case of complete deviation is an indication of the absence of a normal distribution of the data, as a result of all or most of the answers not gathering around the center, so it must be known that values matching the normal distribution have been obtained. (Skewness) rule is [15–17].

- The value of (CR) should be between (± 2.58) at a significant percentage (0.01) and between (± 1.96) at a significant percentage (0.05).
- The Z-Value should not exceed <2.
- Z-Value or Z Skewness = statistical ratio over error (standard deviation).

As for the (Kurtosis) test or what is called (the height range), it indicates the high and low normal distribution of the data, and indicates the deviation of the data from the normal center of the answers, and it is either a higher deviation (Peaked) or a lower deviation (Flatness) from the center of the answers, where the normal distribution is ideal at point (0), and the deviation is high when all the answers are up, and the deviation is low when all the answers are down, so it must be known that values matching the normal distribution have been obtained according to the (Kurtosis) rule, which is explained as follows [15–17]:

- The value of (CR) should be between (± 2.58) at a significant percentage (0.01) and between (± 1.96) at a significant percentage (0.05).
- The value of Z-Value should not exceed <7.

- Z-Value or Z Kurtosis = statistical percentage over error (standard deviation).

If all values fall within the normal distribution limits of (±1.96), this indicates that all study data follow the normal distribution.

4.4.1 *Multicollinearity*

The researcher should always work to avoid unreliable results. Because this leads to the weakness of his study in general and the inability to interpret the results in particular. Multicollinearity is one of the statistical indicators that express the weakness of the study results [18]. It expresses the statistical relationship between the independent variables. The value of the statistical correlations between the independent variables should be at very low limits and preferably non-existent. This result usually appears when evaluating the Measurement Models. If the correlation coefficient between the independent variables is ±1.0, they are considered parallel and completely correlated. Which leads to unreliable scientific results. This usually happens if the data for the same indicator is entered twice by mistake or if the data entered for these two variables are very similar, for example, sales in thousands of units and sales in units. Here, the analysis is not feasible in a single matrix, especially using PLS-SEM. Another reason for the emergence of high correlations between variables is the use of redundant indicators in the study model. For example, using more than one indicator to measure a single issue. Therefore, redundant or less important indicators should be eliminated. Multicollinearity affects the estimation of standard weights, whose values are of great importance for expressing the results of the analysis. Here, the standard errors are very high. Therefore, most statisticians recommend that the value of multicollinearity be between 0.01–0.03 [19].

4.4.2 *Common Method Bias*

Response bias means the discrepancy between the opinions or suggestions and answers given by respondents [20]. Response bias usually represents a state of ambiguity in obtaining a clear response to the questions asked. This is attributed to many reasons, some of which are related to the questions or items asked and some of which are related to the respondent. As for the questions asked, there may sometimes be ambiguity or similarity in the scale's items. As for the respondent, the matter may be attributed to psychological, social and functional factors that lead to a lack of consistency in the response between respondents, and thus bias occurs in opinions or answers. Recently, the delay in responding by respondents is one of the elements of bias [21]. The discrepancy in the timing of delivering the response is a matter of great importance for researchers seeking responses at specific and known times.

One of the most commonly used tests to detect common bias is the Harman one-factor test. This test checks whether there is a single factor that explains 50% or less of the total variance in the data [22]. SPSS.V23 can be used to detect this. Smart PLS can also be used. The danger of common bias in the data lies in the fact that it leads to providing an alternative explanation for the assumed relationships between variables according to scientific theory [23].

Recently, social science studies have settled on using second-generation techniques. Due to their advantages that have overcome the disadvantages of first-generation techniques. Also, as we have shown, research models have developed and expanded significantly. The research model consists of several variables. Which prompts the researcher to choose multivariate analysis. So, there are a set of considerations that the researcher must pay attention to when using multivariate analysis, especially if he relies on second-generation techniques [10]:

When the obtains various answers. The study needs to identify the composite value of the answers for each variable. This is called the composite variable [24]. It is considered a combination of several variables that are calculated with weights. To extract the composite value of the variable, the following equation is used:

$$\text{Composite Value} = w1 \cdot x1 + w2 \cdot x2 + \ldots + w7 \cdot x7$$

The value of x represents the questions in the questionnaire, which are called (individual variables). The value of w represents the weights given to each individual variable. The values of x are collected, and then the values of w are collected and multiplied together. As in the equation above, which consists of seven individual variables and seven weights.

4.4.3 Measurement

The term measurement means a lot in our lives. A tailor cannot make clothes if does not use a tool of measurement, which is the meter. A baker cannot make a good cake if does not use a tool of measurement, which is weight. An electrical engineer cannot make a balanced electrical circuit if does not use a tool of measurement, which is the ampere. And so on for the rest of the usual things in our daily lives. So, measurement is the process of specifying numerical symbols that change according to the nature of the thing to be measured. And they are specified according to scientific rules that are consistent with the nature of the thing to be measured. There are things that are clear and can be easily measured based on the approved unit of measurement. As in the previous examples, length, weight, or current intensity can be measured easily and accurately based on the tools used in the measurement. For example, a scale is used to measure weight. A measuring tape is used to measure cloth. An ammeter is used by an electrician to measure current intensity. And so on for other things such as a thermometer used by a doctor to measure a patient's temperature. The previous cases are called observable and measurable variables. They are apparent variables, not latent ones. But in the social sciences, the matter is completely different. There is no fixed unit of measurement that is used by everyone. For example, when you want to measure the level of trust among a group of employees in the company's management. The measurement here is more difficult because the phenomenon to be measured is invisible and intangible and therefore cannot be easily measured. So, the variables of this type of phenomenon are called latent variables or unobservable variables. This type of variable is usually measured using indirect measurement methods. Through a set of alternative measurement units. Which are employed to measure only one concept. For example, if we want to measure the level of trust a customer has in an electrical device produced by Company *X*. We ask the customer the following questions (these questions serve as alternative measurement units). Do you need to look for other customers' reactions? Do you hesitate a lot before buying or do you buy directly? Do you ask the seller for guarantees about the device you want to buy? Many statisticians, such as Hair et al. [19], believe that adopting multiple individual variables (directing a number of questions) to measure one concept (trust) makes the scale more accurate. In addition, when the individual variables are multiple, they will represent the concept (trust) in its various aspects. This leads to reducing measurement error. It is the result of subtracting the true value of the variable and the value obtained through measurement.

In order to obtain an accurate measurement of the data obtained, there are four measurement scales. Nominal, ordinal, interval, and ratio. Each of these four scales expresses a different order of measurement. Nominal is the least used at the present time. Because it is limited in terms of the types of analysis that can be used. In the nominal scale, the description of the situation is more important than the numbers. For example, if we want to measure the sales volume of a product during the four seasons of the year. The sales volume is coded as follows. Spring = middle, summer = high, autumn = middle, winter = low. Here the numbers have no value and the only reliance will be on the description. Ordinal means that the numbers are of utmost importance in describing the phenomenon studied. If they increase or decrease, this means that the value of the variable has increased or decreased. If we code the quality of the product after presenting it to the specialized experts. Low = 0, medium = 1, high = 2. The change in the values means that the level of quality increases and decreases with the increase and decrease in those values. Therefore, the ordinal scale provides us with clear information about the observations that can be explained about the quality of the product. But at the same time there are differences in the level of spacing between values. It cannot be observed that the differences between low quality and medium quality are the same as the differences between medium quality and high quality. Thus, it is not reasonable to calculate the arithmetic mean and standard deviation for them [10]. The Interval scale represents a set of values with significant differences. For example, when a doctor measures the weights of diabetic patients on a daily basis, will get different weights (observations). There are different weights 80, 90, … etc. The difference between 80 and 90 is exactly 10. It is the same as the difference between 90 and 80. The difference between the two weights is called spacing. It is sometimes necessary. The doctor needs to determine the ideal weight for diabetic patients. On the other hand, the Interval scale is useful in determining the zero point of the measurement. This helps many analysts with the appropriate limits for the measurement, such as the temperature of water between freezing 0 and boiling 100. The Ratio scale is a variable quantitative scale. It helps the researcher compare the relative differences of phenomena. The unique feature of this scale is that it starts from zero. A common example of a ratio scale is measuring age groups. 0 to 20, 20 to 40, … etc. It is also used to measure weights, volumes, and time.

In order to obtain a clear and accurate format for the data, the researcher must code before conducting the analysis process. Coding means the process of converting data from descriptive data to numerical data. This makes it easy to organize and perform various mathematical operations on it. Coding is usually used in open-ended questions. It is also used in interviews. In addition, it can be used in surveys and questionnaires. Coding is an important function for analyzing multivariate studies. Because it determines the nature of the intervals associated with the scale. For example, when using a five-point Likert scale, 1 = strongly disagree and 5 = strongly agree, and the rest of the numbers are the options in between. This means that coding determines the interval between the responses to the latent variables to be analyzed. This facilitates the clarity of the distance between the answer intervals.

Data distribution is a graphical representation of the data collected from the study population/sample. When researchers collect data from a study sample, they seek to identify its graphical form. This means identifying the trend of the responses of the targeted population or sample. For example, when a researcher conducts a study on the level of public confidence in the banking services of a bank. There are answers that will be collected and emptied. We assume that these answers were collected using a decimal scale. So, they are distributed on the basis of categories (1, 2, 3, 4, 5, … 10) and displayed in a table. When the answer corresponding to the number 5 is repeated at a higher level, most of the data will be centered around the middle of the number line. Therefore, we can say that the data is normally distributed. However, if most of the answers corresponding to the number 1 are repeated, the data will be centered around the right of the number lines, and therefore we can say that the data is not normally distributed. The same applies if the opposite happens and the data is centered around the number 10. We benefit from knowing the nature of the data distribution in determining the appropriate statistical method. If the data follows a normal distribution, we can adopt parametric statistics. If the data follow a non-normal distribution, non-parametric statistics can be used. Successful assessments of the distribution of data include the Kolmogorov–Smirnov test and the Shapiro–Wilk test [25]. Skewness and kurtosis can also be examined. This allows assessing the extent to which the data deviate from the normal distribution [26].

4.5 Structural Equation Modeling with Partial Least Squares

Path Models are descriptive diagrams that display the form of relationships between study variables. On the basis of them, hypotheses are estimated and tested when conducting the analysis (Celsi et al. 2016). The path model is built based on the theories adopted in the study. The theory is the scientific basis for forming relationships between study variables. Then, the hypotheses are developed individually. There are two main components of the PLS path model. First, the internal structural model. The task of this model in the scope of PLS SEM is to display the variables (unobserved structures) and the relationships (paths) between them. Second, measurement models for internal and external latent variables. They are also called (external models in PLS-SEM). In the case of measuring external latent variables, they measure the elements that explain the other elements in the model. In the case of measuring internal latent variables, they measure the elements that were explained in the model.

Path Models in PLS-SEM Are Represented as Follows:

- The ovals in the path model indicate variables that are not directly measured. These are called (oversight data).
- The rectangles connected to the ovals indicate the oversight variables. Or the proxy variables that contain the raw data. These are called (indicators).
- The single-headed arrows connecting the ovals and the rectangles indicate directional relationships. These are an expression of theoretically supported predictive relationships
- The path model is based on two theories. Measurement theory and structure theory. The first specifies how each structure is measured. The second theory measures the association of the structures with each other.

4.5.1 Assessment of Measurements Model

The task of measurement theory is to determine how to measure latent variables or unobservable variables. This theory is based on two methods of measuring latent variables. The first method is called reflective measurement. In this method, the arrows go from the latent variables to the indicator variables (measurable variables). This indicates the presence of a causal or predictive relationship in this direction. The other method is formative measurement. Here, the arrows go from the indicator variables to the latent variables. This is the exact opposite of the first method. Formative measurement does not show an error index for the indicator variables and therefore does not assume errors in the predictive relationship. While in the reflective measurement method, errors appear in the measurement model [27]. One of the most important tasks that the researcher must perform is to assess the cogency of the equations, which is called the Assessment of Measurements Model. Based on some statistical indicators, the researcher can identify the validity and reliability of the study scale. Because the study scale is the tool based on which data is collected from the study sample. Depend on the answers included in the study scale, the results are obtained and then interpreted [28]. There are six steps that the researcher takes to identify the reliability and validity of the study scale. Three of them are called conversion validity, which are, factor loading (its value must be greater than 0.70), average variance extracted (AVE) test (its value must be greater than 0.50), and finally consistency reliability (CR) (its value must be greater than 0.70). The other three are called discriminant validity, which include the Fornell-Larcker index (the main correlation of the variable must be greater than the sub correlations and its value is equal to the square of the AVE value), the heterotrait-monotrait (HTMT) ratio (its acceptable value ranges from 0.85 to 0.90), and the cross loading index (the values of the paragraphs of the variable must be greater than the rest of the values in the same column of the variable). If the researcher finds that all the values of the required indicators are within the specified range, he can adopt the scale to test the hypotheses. It is worth noting that it is not necessary to conduct all the tests that we have mentioned. It is possible to use two or three tests, according to what statisticians have indicated [29, 30].

4.5.2 Assessment of Structural Model

The structural theory explains the form of the association of latent variables with each other. In other words, it explains the form of the path model of the relationships between latent variables. Latent variables are the variables that are explained in the model. According to the scientific theory, there are relationships between these latent variables. The process of drawing the path model starts from left to right. of course, the process starts with the independent variable/variables. Then the dependent variable/variables, which can be dependent and mediating at the same time. The oval shapes from which the arrows start are independent variables. Which are usually called exogenous latent variables. As for the ones that the arrows enter and exit, they have mediating variables. In contrast, the dependent variables are those that only the arrows enter. These two types of variables are called endogenous latent variables [28]. After the researcher has finished evaluating the measurement model and the statistical indicators appear acceptable, moves towards evaluating the structure of the model. The concept of Assessment of Structural Model refers to evaluating the study model, in other words, evaluating the path of relationships established by the researcher depend on the scientific theory [31]. Therefore, the results of the causal relationships between the variables of his study, which are (i.e., Independent, moderate, mediating, control, and dependent variables). At this stage, the R^2 value for the latent variables is evaluated, the path coefficients are estimated, the effect size (F^2), and the importance of prediction Q^2. As for the coefficient of determination (R^2), it is an expression of the size of the variance in the dependent variable that can be predicted through the effect of the independent variable. Since PLS-SEM describes the latent internal variance, the value of (R^2) is of great importance in expressing the size of the variance [31]. Estimating path coefficients or testing hypotheses. Bootstrapping is the best method for estimating path coefficients. Because through it, a large number of subsamples are taken to provide standard deviations expressing the estimated (t) value [10]. To estimate the level of influence of the independent variable on the dependent variable, (F^2) is adequate in the multiple regression model. Finally, the value of the predictive relevance indicates the level of accuracy in the hypothesized model and its ability to reconstruct the values and estimate the obtained parameters [28].

4.6 Bibliographic Analysis

Studies that have adopted a dual-phase SEM and ANN analysis consume augmented very. In step with such studies is problematic as there are enormous aids and watercourses of investigate on SEM and ANN models. Scholars utilise a variety of qualitative and quantitative work appraisal methods to comprehend and systematise previous results. Bibliometrics can deliver a transparent review process based on the arithmetical techniques used for research, scientists, or technical activities [32]. Adopting this examination tool is valuable in keeping pace with, charting, analyzing, and imagining information from earlier literature. In this setion, we have done an in-complexity examination of the amount of articles by grouping and database, yearly scientific invention, country scientific production, trend themes, factorial study and country teamwork network to hut light on SEM and ANN models and deliver an impression of such subjects for ongoing investigate, as shown in the following sub-sections.

As shown in Fig. 4.1 the highest number of relevant articles selected from is attributed to the social science scholars who managed to publish in WOS indexed journals. Whereas the fewer number of articles selected from Scopus and IEE Xplore refers to the limited number of journals in these databases which publish social science research. Outcomes settle that the articles on the SEM and ANN model are increasing fast. The databases matched the available articles to enhance the competence of knowledge ended earlier articles on SEM and ANN. In this situation, journals in ScienceDirect and WoS have taken an attention in SEM-ANN studies. Therefore, Scopus and IEEE should appreciate the attention to the new tendency in SEM-ANN studies and should provide more publication space. Journals can be encouraged to publish special issue[s] on SEM-ANN techniques for further discussion and improvement in the method and to settle the cogency of the SEM results.

Figure 4.2 displays the annual trends in of research articles published between 2006 and 2025 in the already mentioned two categories. The figure shows that there has been a significant increase in number of research articles publication during the year 2024.

Similarly, in the year 2020, there was highest number of theoretical based publications as compared to rest of the years during the mentioned chosen period. There is also highest number of studies relevant to technology cases in 2020 as compared to similar studies in other years (except 2022) and this is due to the reason of increase

Fig. 4.1 Documents overview

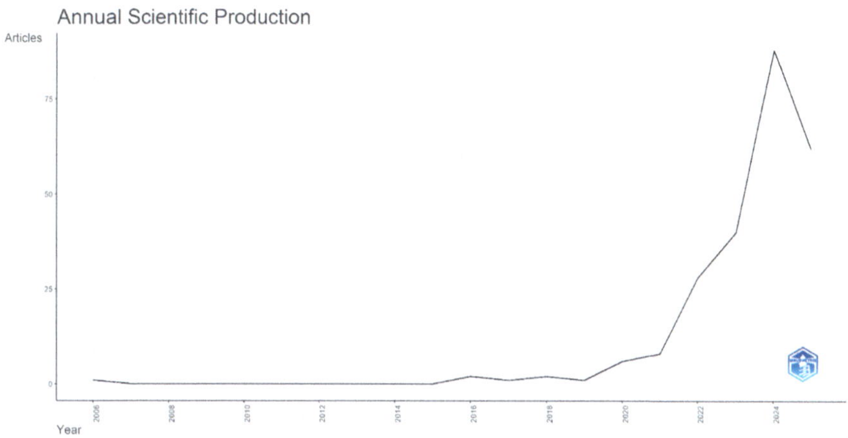

Fig. 4.2 Yearly scientific inventions

in issues corresponding to technology during the recent times. These studies are followed by those published in 2022, which focused on marketing cases and the topics about customer intentions towards online purchase.

Maximum of the literature before 2020 accepted the application of the ANN approach based on an automatic building. Moreover, the popularity of the works that combined SEM and ANN models has applied the sensitivity method techniques to calculate the significance of the standards. Hence, Fig. 4.3 displays the key donors to methodical production, conference articles and scholarly articles. Methodical articles have shown an enormous interest in topics that involve mutual SEM and ANN analysis. Methodical production amplified from ten to fourteen between 2021 and the beginning of 2022. So, yearly production designates a hopeful new trend towards utilizing a dual-stage SEM and ANN analysis. The actuality of such unresolved scientific issues shows SEM and ANN analysis will similarly develop in the upcoming. Furthermore, this rise in data is a sign of incessant growth, which should be experiential by researchers, as well as its pertinence in connected arenas.

Tendency subjects are those that are currently under discussion in a certain field. They signify ideas and subjects analysed in the earlier literature for the SEM and

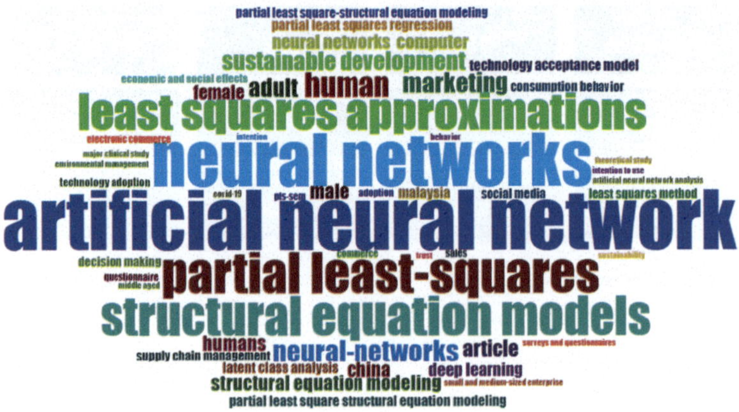

Fig. 4.3 Trend topics

ANN models. Figure 4.3 reflect the major and important key words chosen from the earlier published research on SEM-ANN.

Trend topics help to understand the research taxonomy which exemplifies and substantiates that maximum projects which joint the SEM-ANN absorbed on two issues. For example, Fig. 4.3 shows that the mostly used keyword is 'technology' (cloud computing and technology adoption), shadowed by those used in the advertising issues that adopted mobile buying or spending perception, as well as mobile trade. The keywords used by the authors reflect their other research interest in addition to SEM-ANN. Trend subjects can provision the search position by descriptive the ratio of keywords. Figure 4.3 displays the most important factors of text data which were organized. This process banned scholars and receivers comparation the various mechanisms since discovery the resemblances and changes between the greater and braver words. Also, trend subjects help scholars and practitioners to appreciate the meaning of SEM- ANN acceptance by alliance words of numerous dimensions.

Co-occurrence calculates the similarity index, whilst this analysis allows the user to normalise bibliographic coupling, co-occurrence, and co-citation a similarity measure. Co-occurrence is used to map the conceptual structure of a field based on the frequency of words in a particular bibliographical group. Such an analysis contributes to providing insight to the academics and practitioners into understanding the relationship between the main topic and emerging subfields. The size of nodes shows the frequency of occurrence of keywords. To this end, co-occurrence was used to understand the relationship of the topics which applied SEM with ANN and Fig. 4.4. It explains co-occurrence.

Each bunches are concurrent keywords in the articles existence tested. The red bunch is the main net of keywords comprised the key issue of SEM and ANN models. The works which has adopted a lined and non-compensatory association method has been related with other bulges such as ecological technology, behavioural purposes, cloud computation and implementation of novel knowledge. The blue bunch donates to providing vision into HR behaviour on societal trade platforms. Moreover, the blue

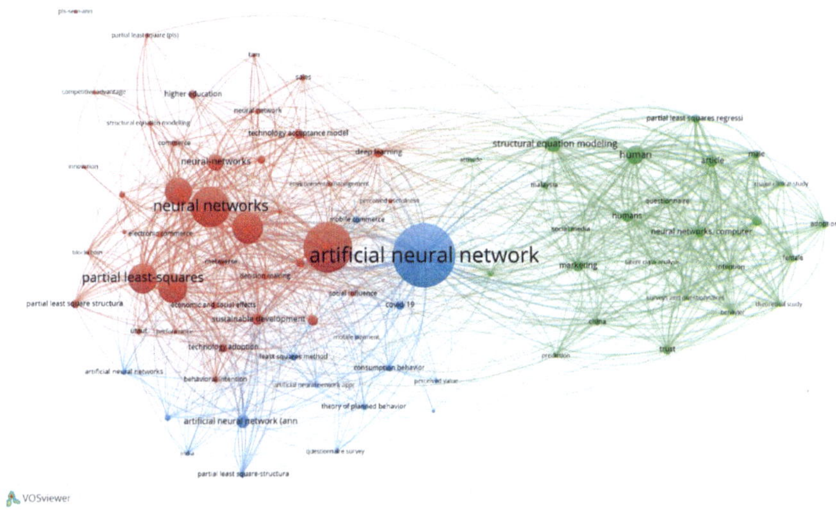

Fig. 4.4 Co-occurrence

bunch is related with sympathetic the clients' trust in societal trade. Examination of factorial study in terms of theoretic and applied based runs the building chunks for continuing investigate and growth in the field of SEM and ANN models.

The knowledge partnership network represents a societal net comprising of bulges and relations. Nodes refer to studies and relations related to co-productions. Co-productions has develop one of the most public kinds of methodical partnership. Scientific collaboration includes the communication of dual or extra scientists in a societal situation to comprehensive errands and attain a mutual area. Flaking light on technical collaboration in SEM and ANN. Figure 4.5. Illustrated nations' partnership.

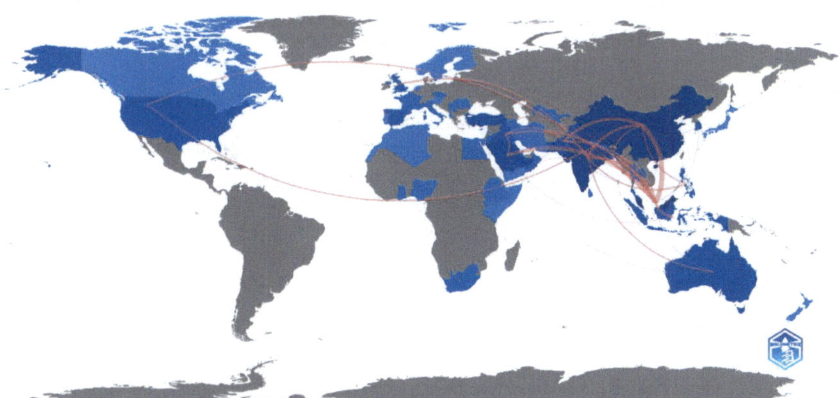

Fig. 4.5 Collaboration world map

Figure 4.5 displays scientific collaboration among countries in a societal context. The red linked appearances designate the existence of obliging relatives among an agreed couple of countries. The bushier the edges between the lines linking countries, the sturdier the obliging relations would be. Moreover, China and Malaysia are the essential countries for scientific collaboration in SEM and ANN models. Such players produce a large amount of research literature in the network of scientific collaboration. Players in dim blue colour are the most careful and active in publishing. However, the light blue colours reflect countries which made fewer contributions to the publication.

References

1. Adams, R.H., Jr., Page, J.: Poverty, inequality and growth in selected Middle East and North Africa countries, 1980–2000. World Dev. **31**(12), 2027–2048 (2003)
2. Fornell, C., Zinkhan, G. M.: Classification schemes for a new generation of multivariate analysis. Marketing Theory, Philosophy of Science Perspectives, 295 (1982)
3. Hair, J.F., Sarstedt, M., Ringle, C.M., Mena, J.A.: An assessment of the use of partial least squares structural equation modeling in marketing research. J. Acad. Mark. Sci. **40**, 414–433 (2012)
4. Chin, W. W.: Commentary: issues and opinion on structural equation modeling. MIS quarterly, vii-xvi (1998)
5. Sarstedt, M., Ringle, C.M., Henseler, J., Hair, J.F.: On the emancipation of PLS-SEM: a commentary on Rigdon (2012). Long Range Plan. **47**(3), 154–160 (2014)
6. Henseler, J., Dijkstra, T.K., Sarstedt, M., Ringle, C.M., Diamantopoulos, A., Straub, D.W., Calantone, R.J., et al.: Common beliefs and reality about PLS: comments on Rönkkö and Evermann (2013). Organ. Res. Methods **17**(2), 182–209 (2014)
7. Bobee, B., Cavadias, G., Ashkar, F., Bernier, J., Rasmussen, P.: Towards a systematic approach to comparing distributions used in flood frequency analysis. J. Hydrol. **142**(1–4), 121–136 (1993)
8. Song, S., Zhang, A., Wang, J., Yu, P.S.: SCREEN: stream data cleaning under speed constraints. In Proceedings of the 2015 ACM SIGMOD International Conference on Management of Data, pp. 827–841 (2015)
9. Hosking, J.R.M., Wallis, J.R.: Some statistics useful in regional frequency analysis. Water Resour. Res. **29**(2), 271–281 (1993)
10. Hair, J.F., Jr., Matthews, L.M., Matthews, R.L., Sarstedt, M.: PLS-SEM or CB-SEM: updated guidelines on which method to use. Int. J. Multivariate Data Analys. **1**(2), 107–123 (2017)
11. Mircioiu, C., Atkinson, J.: A comparison of parametric and non-parametric methods applied to a Likert scale. Pharmacy **5**(2), 26 (2017)
12. Shaffer, D.W., Serlin, R.C.: What good are statistics that don't generalize? Educ. Res. **33**(9), 14–25 (2004)
13. Hair, J.F., Gabriel, M., Patel, V.: AMOS covariance-based structural equation modeling (CB-SEM): guidelines on its application as a marketing research tool. Brazilian J. Market. **13**(2) (2014)
14. Blumberg, B., Cooper, D., Schindler, P.: EBOOK: Business research methods. McGraw Hill (2014)
15. Arnau, J., Bendayan, R., Blanca, M.J., Bono, R.: The effect of skewness and kurtosis on the robustness of linear mixed models. Behav. Res. Methods **45**, 873–879 (2013)
16. Bono, R., Arnau, J., Alarcón, R., Blanca, M.J.: Bias, precision, and accuracy of skewness and kurtosis estimators for frequently used continuous distributions. Symmetry **12**(1), 19 (2019)

17. Cain, M.K., Zhang, Z., Yuan, K.H.: Univariate and multivariate skewness and kurtosis for measuring nonnormality: prevalence, influence and estimation. Behav. Res. Methods **49**, 1716–1735 (2017)
18. Sarstedt, M., Hair, J.F., Ringle, C.M., Thiele, K.O., Gudergan, S.P.: Estimation issues with PLS and CBSEM: Where the bias lies! J. Bus. Res. **69**(10), 3998–4010 (2016)
19. Hair, J.F., Jr., Sarstedt, M., Matthews, L.M., Ringle, C.M.: Identifying and treating unobserved heterogeneity with FIMIX-PLS: part I–method. Eur. Bus. Rev. **28**(1), 63–76 (2016)
20. Bell, E., Bryman, A., Harley, B.: Business Research Methods. Oxford University Press (2022)
21. Lindner, J.R., Wingenbach, G.J.: Communicating the handling of nonresponse error in Journal of Extension Research in Brief articles. J. Extension **40**(6), 9 (2002)
22. Fuller, C.M., Simmering, M.J., Atinc, G., Atinc, Y., Babin, B.J.: Common methods variance detection in business research. J. Bus. Res. **69**(8), 3192–3198 (2016)
23. Bagozzi, R.P., Yi, Y., Singh, S.: On the use of structural equation models in experimental designs: two extensions. Int. J. Res. Mark. **8**(2), 125–140 (1991)
24. Grace, J.B., Anderson, T.M., Olff, H., Scheiner, S.M.: On the specification of structural equation models for ecological systems. Ecol. Monogr. **80**(1), 67–87 (2010)
25. Sarstedt, M., Mooi, E.: A Concise Guide to Market Research, vol. 324. Springer, Berlin (2014)
26. Hair, J.F., Risher, J.J., Sarstedt, M., Ringle, C.M.: When to use and how to report the results of PLS-SEM. Eur. Bus. Rev. **31**(1), 2–24 (2019)
27. Diamantopoulos, A.: Incorporating formative measures into covariance-based structural equation models. MIS Quarterly, 335–358 (2011)
28. Hair, J.F., Jr., Howard, M.C., Nitzl, C.: Assessing measurement model quality in PLS-SEM using confirmatory composite analysis. J. Bus. Res. **109**, 101–110 (2020)
29. Ali, F., Rasoolimanesh, S.M., Sarstedt, M., Ringle, C.M., Ryu, K.: An assessment of the use of partial least squares structural equation modeling (PLS-SEM) in hospitality research. Int. J. Contemp. Hosp. Manag. **30**(1), 514–538 (2018)
30. Shmueli, G., Sarstedt, M., Hair, J.F., Cheah, J.H., Ting, H., Vaithilingam, S., Ringle, C.M.: Predictive model assessment in PLS-SEM: guidelines for using PLSpredict. Eur. J. Mark. **53**(11), 2322–2347 (2019)
31. Henseler, J., Ringle, C.M., Sarstedt, M.: A new criterion for assessing discriminant validity in variance-based structural equation modeling. J. Acad. Mark. Sci. **43**, 115–135 (2015)
32. Aria, M., Cuccurullo, C.: Bibliometrix: an R-tool for comprehensive science mapping analysis. J. Informetr. **11**(4), 959–975 (2017)

Chapter 5
Opportunity for ANN Analysis

Abstract This chapter highlights the concept of electronic word of mouth. Research on e-WOM has expanded significantly, especially over the past decade. The chapter demonstrates the impact of e-WOM on changing consumer behavior. Both positive and negative word of mouth reshape consumer behavior. Perceived crowding plays a significant role in enhancing brand awareness and demonstrating brand credibility. The chapter also includes an understanding of technology acceptance to activate psychological motivations. Furthermore, the chapter addresses the importance of positive feedback in establishing trust in the manufacturer and, consequently, the adoption of online platforms. Conversely, when feedback is negative, it leads consumers to reconsider the marketing channel.

Learning Objectives

1. This chapter aims to understand e-WOM across online platforms and its role in reshaping consumer behavior.
2. Readers understand how the impact of electronic words is amplified when individuals' reactions interact in the digital environment.
3. This chapter aims to clarify the relationship between perceived crowding on online platforms and consumer decision-making.

5.1 ANN Applications

In recent times, businesses have heavily relied on the internet and social media platforms as an important avenue for advertising purposes. Internet and other social media platforms facilitate businesses to instantly evaluate customers' reactions [s] towards their products and services. Customers convey their reactions through reviews, vlogs, rankings, etc., commonly known as electronic word of mouth (eWOM). There has been numerous evidence that e-WOM is quite significant to influence both existing and potential customers in terms of their decision to purchase online products and

X. Chew et al., *Partial Least Squares Structural Equation Modeling and Complementary Methods in Business Research*, Information Systems Engineering and Management 67, https://doi.org/10.1007/978-3-032-01055-1_5

services [1]. Customers can give positive or negative reviews, which constitute positive and negative eWOM.

5.1.1 *Nature of Online Information/Electronic Word of Mouth (eWOM)*

Information technology has revolutionized the ways of doing business with instant access to the market. Companies can easily obtain the market information and judge the customers' reactions [s] to respond quickly. The development of social media platforms has curtailed the market barriers and facilitated the instant communication between vendors and customers [2–5]. With the help of the internet and various social media platforms, companies can quickly reach customers and can quickly revise their marketing campaign to align with the market needs. There are numerous other advantages of using information technology, such as instant delivery of services, reduction in cost and time saving, customization of products and services, and reduced delivery time [6]. It is not only firms, but customers can equally benefit from the advent of information technology and other social media platforms. Now, customers can easily convey to companies and other potential customers their feelings and experiences with products and services. Social media platform facilitates customers in terms of information sharing and feedback, commonly called as electronic word of mouth eWoM (Lim, 2020) defined "any positive or negative statement made by the existing and potential customers about a product, service, or company, which is made available to a multitude of people on the internet; for example, this can include online communication in social networking sites, forums, and consumer review sites". E-WOM can be considered a narrative formed through online communication between buyers on different social media platforms. Availability of online reviews EWOM helps customers (Smith, 2021) make informed purchase decisions. Social media platforms accommodate several customers, which constitutes an online community of customers. They are largely dependent on the availability of online information for making online purchase decisions [s] [7]. Various surveys have forecasted that there will be 266.7 million online customers in the USA by the end of 2022, and this number will further reach 291.2 million by the end of 2025. Similarly, these surveys have also concluded that in the future, there will be an increase in online purchases across the world. This indicates that the use of information technology for business purposes has enriched customers' online shopping experiences through the instant provision of a variety of customized products and services. This significant increase in online purchases is credited to the availability of relevant online information called e-WOM. E-WOM encourages potential customers to broaden their online search of products and services and to choose the best among the variety of available products and services. Companies around the globe are investing huge amounts in the development of their respective information technology infrastructures. For example, in the

US during the year 2021, total investment in information technology was 1.94 trillion US dollars, whereas the total spending for the same was 4.2 trillion US dollars across the world by companies. Today's firms are largely using the internet as a competitive marketing tool to gain and sustain a competitive advantage. This is done through timely obtaining market information and responding accordingly [8–10]. Online information or e-WOM helps companies to evaluate the behavioral intention of customers towards online purchase and thus to redefine their online promotion and other strategies. In today's digital competitive market, it has become mandatory for firms to properly analyze and respond to the online information because this information reflects the behavioral response of customers towards online shopping. In this regard, many research studies can help businesses to this end. These studies are more concerned with explaining the significance of eWOM in influencing online shopping behavior of customers and the ways different social media platforms (i.e., Facebook, Twitter, Instagram, etc.) inform and shape customers' behavior[s] towards online shopping. In addition to an online community of customers, there is Online Group Buying (OGB). OGB is a collective online buying of products and services at reduced prices. These online groups are also dependent on online e-WOM for their decision to make online group purchases. Various studies about online purchases have demonstrated that customers are more concerned about the perceived value and potential risk while making online purchases. This is because they cannot see the product physically. This absence of visibility is an important factor that significantly affects customers' decisions to make online purchases. There is an ample body of research that provides enough evidence of a significant correlation between the availability of online information and customers' perceptions about online available products and services. There is also evidence to the fact that the availability of online information leads to an increase in the perceived value and a reduction in the perceived risk associated with online available products and services. This is due to the reason that more available online information will lead to more informed decisions [s] by the customers. A vast body of literature offers conclusive evidence that e-WOM has a significant effect on the online shopping behavior[s] and attitudes of the customers. Social media platforms provide customers to share their e-WOM in the form of reviews, feedback, and experiences, and thus to make suggestions to other potential customers to help them in deciding whether to go for an online purchase or not. For many potential customers, e-WOM is a source of authentic information that shapes their online shopping attitude and intentions. It is considered a major factor influencing the customers' intentions towards online available products and services. There is a continuous generation of e-WOM on social media platforms and companies' websites because customers not only see e-WOM as a helping guide for online shopping, but they also create e-WOM once they have made an online purchase. e-WOM constitutes a greater source of verified information in the form of subjective experiences shared by the customers; it is a more cost-efficient and reliable source of information, with instant access to relevant and rich information. Usually, customers consider it mandatory to seek the opinion of their peers and other close group members before their a decision to shop online. Based on available information, they evaluate the products through cost–benefit analysis. E-WOM can serve this purpose efficiently and effectively, and hence help customers to make wiser and well-informed decisions. e-WOM helps customers to make their judgment about the pros and cons of online purchases because e-WOM offers a sufficient and relevant information database to make judgments. As it has been described earlier that e-WOM can be both positive and negative. Various studies conducted on the different aspects of e-WOM sufficiently establish that negative e-WOM is more effective than positive e-WOM in influencing potential customers' attitudes and behaviors. This is because there are unhappy and unsatisfied customers are more vocal on online platforms as compared to the satisfied and happy customers. This makes potential customers more cautious in their risk and value perception[s] about the online products and services.

In recent years, the development of internet technology (Web 2.0) and other digital equipment has been rapid and complete transformation in the ways of communication and interaction between vendors and buyers. There was a time when buyers only relied on a single source of information from the vendors. But now they have access to wider and much-improved information sources. In this regard, social media platforms store a huge amount of information that can be easily and instantly accessed by customers. Social media platforms not only constitute an information source but also offer platforms [s] to customers to generate information to be used by other potential customers and firms [11]. The existing customers significantly affect the opinions, attitudes, and online shopping behavior of potential customers and likewise influence the vendors to revise the policies and strategies concerning online products and services.

The exchange of product[s] and service[s] related information through social media platforms between the existing and potential customers is called e-WOM. It is mostly in the form of videos, snapshots, written feedback, audio, and video recordings. It is a way through which existing customers give their subjective views after they have consumed the product[s] and service[s]. This constitutes an unbiased review to guide potential customers in their decision-making.

e-WOM has three distinguished features that affect the potential customer's decision-making regarding online purchases. These are 'opinion seeking', 'opinion giving', and 'opinion passing'. The first component is when potential customers intend to pursue and seek out the required information provided by the existing customers. Existing customers share their reviews and feedback with potential customers and the companies through social media platforms and other online communication channels. This refers to opinion giving. Potential customers pass on this information to other customers, which is called opinion passing. There are also distinct factors which affect these three components of e-WOM (i.e., seeking, giving, and passing). These factors include information quality, website quality, and information quantity [12]. The first one is the degree of extent to which online available information is useful and effective as perceived by the potential customers while making online purchase decisions. Perception of information quality is dependent upon three factors, i.e., information needs to be novel, understandable, and relevant. To be novel means that information is new and unique and was not available before, understandable means that the customers derive the same meaning from the information that vendors want to communicate, whereas being relevant means that information is aligned with the requirements of customers. Customers will not accept information if they are not novel, understandable, and relevant because they cannot make a conclusive decision[s] if there is the absence of these three factors from the online available information. Another important factor that affects the customer's intentions towards online purchase is referred to as website quality. It means that the website content and information available are rich and satisfy the customers' need for information. To improve the quality of the website, companies need to develop attractive, interactive, information-rich rich and relevant content. The third factor is information quantity, which describes the level and amount of online available relevant information and the extent to which such information is perceived as useful by the potential customers. Ranking of online available information is also an important constituent of e-WOM. Ranked information can help potential customers quickly filter the requisite information. This helps in the quick evaluation of perceived risk and value of online purchases. It is obvious that customers like and intend to purchase products with more positive feedback and are ranked highly by the customers [13]. In this way, information quantity facilitates customers in making cost and time-effective decisions to make online purchases.

5.1.2 Positive and Negative eWOM

Customers' feedback or reviews in the form of eWOM is dependent on their respective experiences with the product[s] and service[s]. If the product[s] and service[s] meet[s] customers' expectations, it results in positive feedback or positive e-WOM and vice versa. Usually, customers give their feedback through the company's website, which asks for customer reviews. The company's website and other social media platforms encourage positive feedback [14]. Some earlier studies demonstrate the positive impact of positive e-WOM on customers' online purchase intentions (e.g., [13, 15, 16]). Ahmad and Laroche [7] described the following components of positive eWOM:

- *Aesthetics*: it refers to the customers' positive feedback on the physical features of a product, like packaging, design, durability, etc. Companies focus more on the physical aspects of products to look more appealing to the customers and generate positive e-WOM.
- *Branding*: Customers are available with online products and services from a variety of vendors. While making an online purchase decision, they make a comparison between the products offered by different vendors, which can be called brand comparison. Usually, renowned brands get more positive e-WOM as compared to new ones.
- *Technical aspects*: Customers also present their e-WOM on the technical aspects of the products. This is largely dependent on how much product-related technical

knowledge customers possess. This is in case of such products with are technical, such as mobile phones, computers, other electronic equipment, etc. Technical aspects of a product mean the product specifications, i.e., speed, capacity, processing, power consumption, etc.

- *Online specific*: This aspect deals with the nature and quality of online information available about products and services. Customers are more interested in locating the relevant online information before making an online purchase decision[s]. There is an online discussion among the customers about the different aspects of a product or service.
- *Helping the company*: This is another major factor in generating positive e-WOM. Customers can give positive reviews and feedback (positive e-WOM), which inspires other potential customers to decide in favor of an online purchase. Helping the company means securing more customers [17].
- *Affective*: Customers express their affection for certain products and services through positive e-WOM. The use of words like 'enjoy', 'happy', etc., in e-WOM review represents customers' affection with the product[s] or service[s] [18].

Negative e-WOM represents customers' negative experiences with the products and services when products and services do not fulfill the needs of customers. Negative e-WOM also harms the company in terms of negative market reputation [11]. Basuroy et al. [19] argued that negative e-WOM has bad effects on the reputation of a firm over social media platforms which resulting in reduced sales. It has been found that negative e-WOM has negative effects on the online purchase intentions of potential customers [6]. It becomes difficult for firms and social media platforms to bear the consequences of negative reviews. It requires a lot of effort on the part of vendors to change customers' perceptions already influenced by negative e-WOM [20, 21]. It has been found that negative e-WOM has long longer-lasting influence than the positive ones, and it is difficult for firms to overcome the losses caused by negative e-WOM [20, 22]. With the use of Machine Learning Techniques, Ahmad and Laroche [7] found six major components of negative e-WOM. These are described below.

- *Online order anomaly*: It happens when customers are not satisfied with the way their online orders are dealt with. The use of words like annoyed, disappointment, slow processing in reviews represents online order anomalies.
- *Core functionalities*: Customers express their annoyance and disappointment when the product fails to perform its basic functions.
- *Mishandling*: This relates to the inappropriate and careless handling of the products and services on the part of the vendors.
- *Warning*: When the customers are not satisfied with the products and services, they inform potential customers through a warning message about the purchase and usage of the product or service.
- *Value*: When making an online purchase then customers compare the offered price with the perceived quality of the product or service. Customers perceive lower quality when they realize that the price is higher than the perceived quality. When the potential customers compare the price and quality of the product [23].

- *Shipping charges*: Delayed shipping and deliver along-with higher shipping or delivery charges discourage customers to make online purchases. This consequently results into negative e-WOM.

The earlier studies have found that these are all internal factors and are important to describe the negative eWOM [14, 17, 24].

5.1.3 Perceived Crowding

With the development of digital technologies and social media forums, customers are no more required to visit the markets physically to find the products and services of their choice. Now they can find a large collection variety of products and services which meet their needs. In the same way vendors have found social media platforms and website as convenient way to offer products and services. Though it seems convenient both for vendors and customers but there is a potential issue of choosing the product[s]. Many online available products and services offer similar advantage[s] with no or little difference is prices. Customers get confused to select the one over the other. Social media platforms also connect existing and potential customers to form a community of online customers. They have common interests and are likely to seek the opinions of other community members to make online purchases. Lim [25] presents theory of 'crowding' in study of impact of Covid-19 on tourism industry. Lim described the concepts of 'overcrowding and under crowding and stated that tourism places are either overcrowded or undercrowded. Crowding refers to the total number of people at a tourist place at a certain time. The concept of crowding is equally applicable to the presence of online community of customers on different social media platforms. Some social media platforms are 'overcrowded' others are 'undercrowded'. This depends on multiple factors, for example, website quality, information quality, perceived value, and risk etc. Lim [25] contended that "crowding beyond a particular density and psychological tolerance is unacceptable".

While elaborating the theory of crowding, Lim described two different concepts of 'agency and reactance' used in studies on tourism. According to Lim, people intend to use authority when it deems ethically right and legal which Lim calls 'agency' theory. The other one reactance theory states that individuals are free to demonstrate behavior of their choice and they resist when they feel resisted to demonstrate their chosen behavior[s]. It is assumed that customers in online community demonstrate such kind of behaviors agency vs reactance as and when seems appropriate. Machleit et al. [26] also narrated the concept of perceived crowding and refers to the density of a particular space. Crowding on a specific destination or space causes to develop and enhance the networks between the individuals and is a source of social support. In the context of social commerce and social media platforms such perceived social support has significant positive influence on the online purchase intentions of the potential customer [27]. McGrew [28] introduced the concepts of spatial and human crowding and are the major factors to influence perceived level of crowding by online customers over social media and other online platforms. McGrew defined spatial crowding in terms of presence of online customers over a certain platform. In the context of social commerce, firms must make enough space available to accommodate more customers and to give their subjective opinions about the online available products and services [29].

Enough provision for spatial crowding allows customers to evaluate the products and form their own opinions about the products and services. Machleit et al., defined human crowding as comments and reactions appearing on social media and other online platforms and the total number of networks created through such comments and reactions. Customers write their comments and show their reactions to reflect their happiness or displeasure with the products and this significantly influence the other potential customers accordingly. Tran [27] argued that negative perception of crowding leads to stress whereas positive perception of crowding results into pleasure and excitement. Spatial crowding helps to increase the total number of customers over online platforms and thus benefits the vendors [30]. Positive perception about the spatial crowding also increases the probability of positive comments about the online products and services on certain online platforms [31]. It is concluded that positive perception of spatial crowding encourages more customers to join online platforms and negative perception of spatial crowding hurdle the entry of more potential customers over the online platforms.

5.2 Customers' Intention of Using New Channels

In addition to other factors (website quality, information quality etc.) the distinctive features of social media platforms are also important determinants of customers' intentions to make online purchases [32, 33]. The perceived social support and social presence are the two distinctive features of social media platforms. It has been found that customers are more intended to use such social media platforms where they receive more social support and feel more social presence [34–36]. Customers are not interested in using social media platforms if there is less social support and fewer information. Social support on social media helps potential and existing customers to develop online networks and feel being connected to a larger community of online customers with similar interests and product preference [37]. Social support on social media platforms is considerably influenced by the presence of positive and negative e-WOM which is also considered as key factors to understand the potential customers' intentions to use social media platforms [7]. These intentions are framed within the context of the reviews and comments given by the existing customers [38]. Positives reviews and feedback encourage potential customers to utilize the social media platforms for online purchase purpose, whereas the negative reviews and feedback discourage potential customers from using such platforms [39, 40].

It is important for social commerce firms to realize the significance of two important factors to encourage customers towards using social media platforms. These are the quality of social media platforms which social commerce firms use to promote their products and the quality of information being presented there. In addition to these two factors, social commerce firms must also pay sufficient attention to the other important aspects of the social media platforms, for example, design and the security features [1, 41]. Potential customers are more inclined towards such social media platforms which offers rich, positive, and quality information about the products and services being offered there. Also, the stringent security features boost potential customers' confidence in social commerce firms [41]. Earlier studies have shown that customers do not make an instant online purchase decisions but there is a complete pre-purchase process (information evaluation, comparison, availability of substitutes etc.) which is significant to customers' utilization of online social media platforms [42]. Each step in pre-purchase process is important because each stage allows potential customers to move closer (or away to their online purchase decision (Cheung and Thadani, 2012. Social commerce platforms must offer such features which leave positive impression in the mind of the potential customers about the product[s] and service[s] [40].

In one of the recent studies, Ghosal et al. [43] investigated the influence of customers' demographics (age, income, education, occupancy area, profession etc.) on their intention[s] to do online purchases. The study concluded that there are some important external factors which influence customers' intentions towards online purchases. These factors include ease of doing online purchases, security features of social commerce platforms, delivery time and process, market reputation of the firm and quality of customers' services. The study also found that out of the mentioned demographics, education, and occupancy area, are two major demographics which are relevant to customers' decision of online purchases.

Social commerce platforms and other digital media platforms are highly dependent on the quality of internet for uninterrupted services. Though internet has facilitated the communication between remotely located customers and vendors. Still the absence of physical interaction between the two is a concern on the part of both customers and vendors. Accordingly, potential customers are highly dependent upon the experiences of the existing customers to make online purchase decision [44]. Therefore, customers become reluctant to totally depend upon the information, ratings, reviews, and referrals to build their trust towards unknown vendors over social commerce platforms [45, 46]. Trust is an important and foremost component of online transactions over the social commerce platforms. It significantly influences the level of perceived risk associated with online purchases. Potential customers are highly concerned about the level of trust being shown by the other existing customers towards online vendors [47]. Trust is important factor to influence customers in a way that they feel more secured while using social commerce platforms [48]. Trust is a mutual concept and it takes place between individuals (customers) where one anticipates that the experience and opinions of others is more reliable, valuable, and trustworthy. It is emotion related concept and denotes customers emotion-derive trust towards online vendors [49]. Trust is considered to be an instrumental in mitigating the perceived risk in online transactions. Vendors make their utmost efforts to build customers trust in social commerce platforms by offering the recent reviews and quality information over there [50]. In the era of m-commerce and s-commerce, customers are available with multiple information sources. The growth and development of internet technology it has become very easy to access commercial information provided by the numerous vendors on social commerce platforms. In such situation, it becomes very important for customers to seek recommendations from the reliable platforms before making online purchase decision[s] [51]. Trust is an important concept to be studied in the context of m-commerce because people share information and develop trust on the basis of shared information. Potential customers seek approval of the existing customers while using m-commerce platforms [52]. Trust theory has been used in behavioral studies in social sciences and is considered to be equally applicable in m-commerce studies [53, 54].

References

1. Wang, C.L., Ye, L.R., Zhang, Y., Nguyen, D.D.: Subscription to fee-based online services: What makes consumer pay for online content? J. Electron. Commer. Res. **6**(4), 301–311 (2005)
2. Alnoor, A.: Human capital dimensions and firm performance, mediating role of knowledge management. Int. J. Bus. Excell. **20**(2), 149–168 (2020)
3. Alnoor, A., Eneizan, B., Makhamreh, H.Z., Rahoma, I.A.: The effect of reverse logistics on sustainable manufacturing. Int. J. Acad. Res. Acc. Finance Manage. Sci. **9**(1), 71–79 (2018)
4. Alnoor, A., Tiberius, V., Atiyah, A.G., Khaw, K.W., Yin, T.S., Chew, X., Abbas, S.: How positive and negative electronic word of mouth (eWOM) affects customers' intention to use social commerce? A dual-stage multi group-SEM and ANN analysis. Int. J. Human-Comput. Interact. **40**(3), 808–837 (2024)
5. Alnoor, A.M., Al-Abrrow, H., Abdullah, H., Abbas, S.: The impact of self-efficacy on employees' ability to accept new technology in an Iraqi University. Glob. Bus. Organ. Excell. **39**(2), 41–50 (2020)
6. Gerdt, S.O., Wagner, E., Schewe, G.: The relationship between sustainability and customer satisfaction in hospitality: an explorative investigation using eWOM as a data source. Tour. Manage. **74**(1), 155–172 (2019)
7. Ahmad, S.N., Laroche, M.: Analyzing electronic word of mouth: a social commerce construct. Int. J. Inf. Manage. **37**(3), 202–213 (2017)
8. Aymen, R.A., Alhamzah, A., Bilal, E.: A multi-level study of influence financial knowledge management small and medium enterprises. Polish J. Manage. Stud. **19**(1), 21–31 (2019)
9. Eneizan, B., Mohammed, A.G., Alnoor, A., Alabboodi, A.S., Enaizan, O.: Customer acceptance of mobile marketing in Jordan: an extended UTAUT2 model with trust and risk factors. Int. J. Eng. Bus. Manag. **11**(11), 184797901988948 (2019)
10. Hadi, A.A., Alnoor, A., Abdullah, H.O.: Socio-technical approach, decision-making environment, and sustainable performance: role of ERP systems. Interdiscip. J. Inf. Knowl. Manag. **13**(2), 397–415 (2018)
11. Chang, H.H., Wu, L.H.: An examination of negative e-WOM adoption: brand commitment as a moderator. Decis. Support Syst. **59**(1), 206–218 (2014)
12. Chopra, I.P., Lim, W.M., Jain, T.: Electronic word of mouth on social networking sites: What inspires travelers to engage in opinion seeking, opinion passing, and opinion giving? Tour. Recreat. Res. **2**(1), 1–14 (2022)
13. Duan, W., Gu, B., Whinston, A.B.: The dynamics of online word-of-mouth and product sales: an empirical investigation of the movie industry. J. Retail. **84**(2), 233–242 (2008)
14. Alalwan, A.A., Algharabat, R.S., Baabdullah, A.M., Rana, N.P., Raman, R., Dwivedi, R., Aljafari, A.: Examining the impact of social commerce dimensions on customers' value cocreation: the mediating effect of social trust. J. Consum. Behav. **18**(6), 431–446 (2019)

15. Goldsmith, R.E., Horowitz, D.: Measuring motivations for online opinionseeking. J. Interact. Advert. **6**(2), 2–14 (2006)
16. Liu, Y.: Word of mouth for movies: its dynamics and impact on box office revenue. J. Mark. **70**(3), 74–89 (2006)
17. Hennig-Thurau, T., Gwinner, K.P., Walsh, G., Gremler, D.D.: Electronic word-of-mouth via consumer-opinion platforms: What motivates consumers to articulate themselves on the internet? J. Interact. Mark. **18**(1), 38–52 (2004)
18. Campbell, C., Pitt, L.F., Parent, M., Berthon, P.R.: Understanding consumer conversations around ads in a Web 2.0 world. J. Advertising **40**(1), 87–102 (2011)
19. Basuroy, S., Chatterjee, S., Ravid, S. A.: How critical are critical reviews? The box office effects of film critics, star power, and budgets. J. Market. **67**(4), 103–117 (2003)
20. Chevalier, J.A., Mayzlin, D.: The effect of word of mouth on sales: online book reviews. J. Mark. Res. **43**(3), 345–354 (2006)
21. Noort, G., Willemsen, L.M.: Online damage control: the effects of proactive versus reactive webcare interventions in consumer-generated and brand-generated platforms. J. Interact. Mark. **26**(3), 131–140 (2012)
22. Chen, Z., Lurie, N.H.: Temporal contiguity and negativity bias in the impact of online word of mouth. J. Mark. Res. **50**(4), 463–476 (2013)
23. Gangold, W., Miller, F., Brockway, G.: Word of mouth communication in a service marketplace. J. Serv. Mark. **13**(1), 73–89 (1999)
24. Jahn, B., Kunz, W.: How to transform consumers into fans of your brand. J. Serv. Manag. **23**(3), 344–361 (2012)
25. Lim, W.M.: Toward an agency and reactance theory of crowding: Insights from COVID-19 and the tourism industry. J. Consum. Behav. **20**(6), 1690–1694 (2021)
26. Machleit, K.A., Eroglu, S.A., Mantel, S.P.: Perceived retail crowding and shopping satisfaction: What modifies this relationship? J. Consum. Psychol. **9**(1), 29–42 (2000)
27. Tran, V.D.: Assessing the relationship between perceived crowding, excitement, stress, satisfaction, and impulse purchase at the retails in Vietnam. Cogent Bus. Manage. **7**(1), 1858514–1858525 (2020)
28. McGrew, P.L.: Social and spatial density effects on spacing behaviour in preschool children. J. Child Psychol. Psychiatry **11**(3), 197–205 (1970)
29. Blut, M., Iyer, G. R.: Consequences of perceived crowding: A meta-analytical perspective. J. Retail. **96**(3), 362–382 (2020)
30. Eroglu, S.A., Machleit, K., Barr, T.F.: Perceived retail crowding and shopping satisfaction: the role of shopping values. J. Bus. Res. **58**(8), 1146–1153 (2005)
31. Mehta, R.: Understanding perceived retail crowding: a critical review and research agenda. J. Retail. Consum. Serv. **20**(6), 642–649 (2013)
32. Gao, G., Greenwood, B.N., Agarwal, R., McCullough, J.S.: Vocal minority and silent majority. MIS Q. **39**(3), 565–589 (2015)
33. Lee, E.J., Shin, S.Y.: When do consumers buy online product reviews? Effects of review quality, product type, and reviewer's photo. Comput. Hum. Behav. **31**(1), 356–366 (2014)
34. Khaw, K.W., Alnoor, A., Al-Abrrow, H., Chew, X., Sadaa, A.M., Abbas, S., Khattak, Z.Z.: Modelling and evaluating trust in mobile commerce: a hybrid three stage fuzzy Delphi, structural equation modeling, and neural network approach. Int. J. Human-Comput. Interact. **38**(16), 1529–1545 (2022)
35. Leong, L.Y., Jaafar, N.I., Ainin, S.: The effects of Facebook browsing and usage intensity on impulse purchase in f-commerce. Comput. Hum. Behav. **78**(1), 160–173 (2018)
36. Leong, L.-Y., Hew, T.-S., Ooi, K.-B., Chong, A.Y.-L.: Predicting the antecedents of trust in social commerce—a hybrid structural equation modeling with neural network approach. J. Bus. Res. **110**, 24–40 (2020)
37. Leong, L.Y., Hew, T.S., Ooi, K.B., Tan, G.W.H.: Predicting actual spending in online group buying—an artificial neural network approach. Electron. Commer. Res. Appl. **38**(11), 100812–100898 (2019)

38. Alnoor, A., Al-Abrrow, H., Al Halbusi, H., Khaw, K.W., Chew, X., Al-Maatoq, M., Alharbi, R.K.: Uncovering the antecedents of trust in social commerce: an application of the non-linear artificial neural network approach. Competitiveness Rev. Int. Bus. J. **32**(3), 492–523 (2022)
39. Bigné-Alcañiz, E., Ruiz-Mafé, C., Aldás-Manzano, J., Sanz-Blas, S.: Influence of online shopping information dependency and innovativeness on internet shopping adoption. Online Inf. Rev. **32**(5), 648–667 (2008)
40. Xiao, Z., Zhang, J., Li, D., Samutachak, B.: The effect of e-WOM on country image and purchase intention: an empirical study on Korean cosmetic products in China. Int. J. Serv. Technol. Manage. **22**(1/2), 18–30 (2016)
41. Comegys, C., Hannula, M., Váisánen, J.: Effects of consumer trust and risk on online purchase decision-making: a comparison of Finnish and United States students. Int. J. Manage. **26**(2), 295–308 (2009)
42. Seock, Y.K., Norton, M.: Attitude toward internet web sites, online information search, and channel choices for purchasing. J. Fashion Market. Manage. Int. J. **11**(4), 571–586 (2007)
43. Ghosal, I., Biswas, D., Prasad, B., Behera, M.P.: Envisioning the impact of online shopping: an antecedent to newfangled social fabric through digital consumerism. Int. J. Adv. Sci. Technol. **29**(8), 1400–1408 (2020)
44. Sharma, S.K., Sharma, M.: Examining the role of trust and quality dimensions in the actual usage of mobile banking services: an empirical investigation. Int. J. Inf. Manage. **44**, 65–75 (2019)
45. Alhamdi, M., Noor, R.M.A.S., Abdulla, M., Alnoor, A., Eneizan, B.: How does financial analysis influence the firm's failure in the Iraqi private sector? J. Soc. Sci. Res. **5**(9), 1321–1328 (2019)
46. Al-Tit, A.A., Omri, A., Hadj, T.B.: The driving factors of the social commerce intention of Saudi Arabia's online communities. Int. J. Eng. Bus. Manage. **12** (2020) 1847979019899746
47. Esmaeili, L. Mutallebi, M. Mardani, S., Golpayegani, S. A. H.: Studying the affecting factors on trust in social commerce (2015). arXiv:1508.04048
48. Lin, X., Wang, X., Hajli, N.: Building e-commerce satisfaction and boosting sales: the role of social commerce trust and its antecedents. Int. J. Electron. Commer. **23**(3), 328–363 (2019)
49. Lin, X., Zhang, D., Li, Y.: Delineating the dimensions of social support on social networking sites and their effects: a comparative model. Comput. Hum. Behav. **58**(1), 421–430 (2016)
50. Beyari, H., Abareshi, A.: The interaction of trust and social influence factors in the social commerce environment. In: International Conference of Reliable Information and Communication Technology, pp. 931–944. Springer, Cham (2018)
51. Chen, J., Shen, X. L., Chen, Z. J.: Understanding social commerce intention: a relational view. In: 2014 47th HI International Conference on SystemSciences, IEEE, pp. 1793–1802 (2014)
52. Alhulail, H., Dick, M., Abareshi, A.: The influence of word of mouth on customer loyalty to social commerce websites: trust as a mediator. In: International Conference of Reliable Information and Communication Technology, pp. 1025–1033. Springer, Cham (2018)
53. Fadhil, S.S., Ismail, R., Alnoor, A.: The influence of soft skills on employability: a case study on technology industry sector in Malaysia. Interdiscip. J. Inf. Knowl. Manag. **16**, 255–283 (2021)
54. Liang, T.P., Turban, E.: Introduction to the special issue social commerce: a research framework for social commerce. Int. J. Electron. Commer. **16**(2), 5–14 (2011)

Chapter 6
Artificial Neural Network and Theories

Abstract This chapter discovers the growing dependence of trades on social media and digital platforms for announcements and customer use, accenting the rising importance of electronic word of mouth (eWOM). It highlights how customer appraisals, scores, and responses can affect others' buying intentions, also positively or negatively. The argument is grounded in some key theoretical models, including Social Exchange Theory and Stimulus Overload Theory, which clarify how people procedure social and environmental stimuli on online platforms. Furthermore, it focuses on Social Theories that investigate the effect of interpersonal dealings and the meaning of relationships online. The chapter additionally contains selected models on behaviors toward technology. These models show an understanding of the facilitators and obstacles of client behavior in assuming or fighting new technologies. The Theory of Planned Behavior and its decomposed are also analyzed to comprehend how moods, social standards, and perceived control affect decision-making in digital environments.

Learning Objectives

1. Comprehend the idea of (eWOM) and its role in determining customer behavior.
2. Classify and describe the main philosophies behind customer behavior in digital platforms.
3. Examine how some social factors affect customer appointments in social commerce.
4. Assess the theoretical models applied to forecast and understand customer acceptance. depending on, or resistance to new technologies in social commerce platforms.

X. Chew et al., *Partial Least Squares Structural Equation Modeling and Complementary Methods in Business Research*, Information Systems Engineering and Management 67, https://doi.org/10.1007/978-3-032-01055-1_6

6.1 Literature Review

In recent times, businesses heavily rely on the internet and social media platforms as an important avenue for advertisement purposes. Internet and other social media platforms facilitate businesses to instantly evaluate customer's reaction[s] towards their products and services. Customers convey their reactions through reviews, vlogs, ranking, etc. commonly known as electronic word of mouth (eWOM). There has been numerous evidence that e-WOM is quite significant to influence both existing and potential customers in terms of their decision to purchase online products and services [1]. Customers can give positive or negative reviews which constitute positive and negative eWOM.

6.2 Underpinning Theories

The theoretic growth of this study is founded on two theories including social exchange theory and stimulus theory. The first one explains the ways customers develop their perceptions about online purchase[s] and the change in customers' intention[s] towards online purchase[s]. After reading the reviews and feedback from existing customers, potential customers may divert to other social commerce platforms. Thus, social exchange theory describes such changes and the construction of e-WOM [2]. Social influence from peers (community of online customers) significantly affects the customer's engagement in e-WOM and the way they evaluate the products and services. Customers are engaged in social exchange interaction[s] with other online customers and emphasize the great importance of online reviews and feedback before deciding to make an online purchase[s] [3]. Naturally, customers may change their intentions towards certain products and services or social commerce platforms after reading the reviews, feedback, and opinions of existing customers. The presence of reviews and feedback on a certain social commerce platform significantly influences customers' intentions towards that online platform [4]. The online interaction between the members of the online community of customers is one major determinant of customers' behaviors towards online purchases of products and services. The main emphasis of social exchange theory is that change in customers' intentions and behavior towards online purchase happens due to the exchange of information and opinions between the online customers and there is no forcible change in customers' intentions and behaviors. In this context, social exchange theory and social influence theory support each other [3].

The other important theoretical aspects that explain the connection between perceived crowding and customers' intent [s] towards online purchases are reflected by stimulus overload theory. Stimulus theory explains how the positive and negative e-WOM influence customers' intention[s] towards using social commerce platforms. Customers' intentions towards social commerce platforms are framed in the context of consensus and disagreement between the customers about certain products and services which appears in the form of reviews and opinions over social commerce platforms [5]. Stimulus theory suggests that various stimuli in an online marketing environment are external to the customers' control and thus customers face behavioral stress to back to the normal marketing environment. The theory is explanatory of the influence of overcrowding on customers' perceptions of and intentions towards social commerce platforms. Any change in customers' perceptions and attitude[s] happens to align with such change[s] in the group's perceptions and attitude to which they belong. Customers' engagement in certain behaviors and actions is the result of their perceptions of the event[s] or change[s] happening in their surroundings [6]. Also, the roles of perceived crowding cannot be overlooked in these relationship[s].

6.2.1 Social Support Theory

Social support is not a novel concept in the field of social sciences. It has been widely studied for theory development and has also been used in several previous studies [7]. However, there is no mutual agreement among the scholars to present a single and conclusive definition of the concept [8]. Cobb [9] defined social support as belonging to a certain social network and one's belief that he/she is an important member of the social network to which he/she belongs. Kim et al. [10] narrated that a person feels socially supported when his/her needs are fulfilled, and he/she has access to the requisite information. Social networks include peers, groups, organizations, institutions, families, and others that provide social support to individuals [11]. It relatively a broader concept that encompasses several other sub-concepts including sub-network, perceived [12], informational [13], tangible [14], emotional [15] and belongingness support [13]. Information support refers to the provision of quality and relevant information to help potential customers in their online purchase decision[s]. Tangible support is financial support or providing the required products and services. Emotional support related to the emotional aspects and feelings of affection, love, care, and belongingness means encouraging potential customers to be involved with other online customers through exchange and interaction processes.

Crocker and Canevello [16] described support network in terms of mental, emotional, and physical support one obtains from peers. This theory argues that one's attitude and behavior are largely dependent on the social support he/she receives. The level of social support one receives is closely linked to the extent one is connected and engaged with his/her surrounding social networks. Individuals seek the fulfillment of their needs by their actual engagement with networks [17]. Sarason et al. [18] discussed three dimensions of support which include relational connectedness, disaggregated social provisions, and a sense of support. Interpersonal connectedness means that one's acceptance of social networks is dependent on the structure of that social network. It further states that the nature of interaction[s] between the members and the way such interaction[s] influence[s] network members is defined by the structure of a social network. But this is not necessary. Some studies argue that one's behavior and attitudes toward the social network are independent of the network structure. Some important factors need consideration to realize the positive outcome[s] of social support [19]. Second dimension of 'disaggregated social provision explains the breakdown of social support and its components structure. It states which an individual may not need the support as a whole but on different occasions, he/she looks for specific support from the network[s]. This also states that social support can only benefit an individual when such support is perceived to be aligned with one's need[s] for social support, for example, physical support is not appreciated when one needs psychological help [20]. The last dimension i.e., sense of asset there must be differentiation must be made between one's perception of social support and the actual support one receives. Studies have shown that the perception of social support is more to influence support outcomes rather than the actual support. The intention to seek social support is closely associated with the perception of the availability of social support [21]. The sense of support explains the way one senses social support. It is quite possible that one does not sense available social support as beneficial to meet need[s] for support. Hupcey [12] suggested that the perception of social help is closely linked with one's perception of 'self", 'the matter of others' and the extent of the network relationships. Social help depends on one's belief about the available social support when needed and is unrelated to the actual support received [22].

6.2.2 Social Presence Theory

Generally, social presence refers to one's perception of 'being there'. With the introduction of social commerce, it is important that customers feel their social presence over social commerce platforms. Scholars in the field of information and digital technology have shown their interest in understanding why and how online customers develop their perception of 'social presence' over social commerce platforms [23]. Social presence theory has its roots in the field of Psychology where symbolic interaction[s] and group communication are considered important to one sense of social presence. Social presence seems to be a concept, that is mediated by the other factors, but it is more like a non-mediated way of communicating. Though, different scholars in the field of Psychology have defined social presence differently, however, the essence of 'being there' is common among all definitions [24]. Arbaugh [25] elaborated on the concept of social presence it is the ability[ies] and the way[s] one communicates with others in his/her surroundings and develops interpersonal relationships by projecting his/her traits in a technology-mediated environment. Other scholars (e.g. [26–28]) define social presence in terms of an interaction between the members to promote group cohesion and a collective sense of community between group members where all members of the group feel supported and closely connected. It is also defined as one's feeling of being with others and the perception of support from others [29]. Studies in the field of contact largely use social presence theory to understand the relevance and usefulness of communication technologies for users.

These studies conceptualize social presence as the extent to which one's presence is important for the effectiveness of interpersonal relationships [30, 31].

Thus, in the context of a technology-mediated virtual environment, social presence aligns with the perception of the importance of the quality of communication channels in various settings [30, 31]. Therefore, individuals feel present through the messages they exchange with others. The concept of intimacy is one of the characteristics of social presence, the familiarity an individual feels when interacting with others via visual or audio communications [32]. Immediacy is also an important characteristic. This concept relates to efficient, immediate communication between parties, which creates an attractiveness to communication, as well as a high degree of fluidity [33] (Inizan et al. 2020).

6.2.3 Unified of Acceptance and Use of Technology (UTAUT)

This theory is based on Venkatesh et al. [34]. They developed a concept that explains the intentions that precede technology adoption as well as the behaviors that follow. This means they established a theoretical framework for the drivers of intention to adopt a new technology. What behaviors will emerge from this intention? This theory was influenced by other previous theories, such as the theory of planned behavior (TPB), technology acceptance (TAM), and the theory of reasoned action (TRA). This theory combined the ideas of these three theories to present a unified model of technology and human behavior [35]. UTAUT attributes an individual's intentions to four main components: facilitating conditions, social influence, effort expectancy, and performance expectancy [34]. Demographic factors are powerful influences on these four factors. This theory has been applied in several fields, including the banking sector. Mensah and Khan [36] demonstrated the importance of UTAUT components in assessing the extent of mobile phone use in banking. The UTAUT model has also been applied to e-government services. It successfully assessed the extent of e-government platform usage [37]. Furthermore, the model has been utilized in several fields, such as virtual reality environments [38, 39]. The diversity of UTAUT adoption demonstrates its importance in analyzing the intentions of potential technology users.

6.2.4 Unified of Acceptance and Use of Technology 2 (UTAUT2)

Modern organizations are increasingly conditional on information technology for the smooth functioning of their routine operations. Rapid technological changes have transformed the ways of doing business and organizations face challenge[s] of technology adoption. Rapid technological changes have transformed the ways of doing business and organizations face challenge[s] of technology adoption. Based on the Theory of Planned Behaviour (TRB) and Technology Model (TAM), Venkatesh [40] developed a model of technology adoption which is called 'Unified Theory of Acceptance and Use of Technology (UTAUT). The model explains the differences in behavior[s] towards the adoption and use of technology. There are different models which discuss the individuals' intention[s] and technology adoption behavior [41] for example, TAM and TRB, Theory of Reasoned Action (TRA). These models explain some of common factors that affect social intentions and technology implementation. UTAUT is different from these in a way that it integrates all the earlier models and gives a sufficient explanation of technology implementation [42].

The UTAUT encompasses the numerous factors which influence users' technology adoption intentions. These factors include performance expectations, the level of effort required, and the social influence. Performance expectation refers to one's expectations that technology adoption and use will help to attain the required performance. Level of effort refers to the magnitude of effort required to use the technology. Social influence denotes the extent to which technology assumption and benefits are impacted by others around the user [43, 44]. These factors have a direct impact on user's attitude and intention[s] to use the technology [45]. UTAUT also accounts for the other factors (user's demographics, age, gender, education, etc.) that strengthen or weaken the effect of the cited factors on the user's intent and use of knowledge.

Recently Venkatesh et al. [34] further suggested ATAUT 2 as an extension to their UTAUT model with the addition of three more factors i.e., hedonistic incentive, price worth, and habit. It is estimated that the earlier UTAUT model explains 70% of the variation in user behavior whereas UTAUT2 explains 74% of variations in user behavior towards technology adoption and usage across the distinctive settings.

6.2.5 Diffusion of Innovation Theory

Diffusion innovation theory explains the ways 'innovations' (technology, new ideas) diffuse into a society. It also applies to the diffusion of innovative products and services. The theory also describes how technology features encourage or discourage technology adoption and usage [46, 47]. The innovation process consists of two major activities: beginning and performance. Diffusion is a procedure through which new technology is delivered to the users over an extended time. The user's decision to accept technology is not an instant one but it is a process consisting of four stages maintenance, dissemination, implementation, and adoption [48–52]. It was Ryan and Gross who initiated the research on technology diffusion inspired by studies in rural sociology. Their first study was on the dispersal of mixed grain in Iowa [53]. Later on studies on 'innovation and diffusion' extended to other fields and there are several studies that support the diffusion of innovation theory [54]. Among others, the work of Rogers [55] is more significant in the development of the diffusion of innovation theory. Rogers stated that there are five types of users depending on their intent toward innovation and technology. These are creators, premature adopters, early adopters, late majority, and laggards. Scott and McGuire [56] mentioned five major factors that influence the speed and the extent of adoption of technology and innovation. These five factors include compatibility, comparable advantage, trialability, intricacy, and observability [57, 58]. The theory states that innovation is not only limited to technology competency in addressing technology-related issues but also involves a collaboration process between the inventors and users. The rate and the speed of technology adoption are dependent on the technology features and social status of the users. Technology inventors are concerned with locating the marketplace needs for innovative offerings and propose the diffusion of technology [59]. The application of dispersal of invention theory is not limited to information technology but is also extended to other fields of sociology, anthropology, geography, public health, economics, etc. [60].

6.2.6 Innovation Resistance Theory

Resistance is a common phenomenon that is faced with every change and innovation. Technology companies face a multitude of challenges owing to resistance towards technology acceptance and adoption. Technology companies are more concerned with understanding the nature of technology resistance and developing ways to diffuse such resistance [61–64]. Innovation resistance theory (IRT) provides a theoretical explanation of confrontation behavior and thus helps in identifying factors of customers' resistance [65]. The theory states that users are at ease with the use of existing technologies and any change is perceived as threatening to their status quo. Thus, resistance is understood as an outcome of the rational understanding of the introduction of new technologies, and deviation from the existing technology is perceived as not aligned with the current needs of users [66]. On the part of the technology innovators, the resistance to technology is seen as a disruption in the current work practices and is deemed necessary to be addressed in a way that users could reply positively to the new skill [67]. Resistance happens when innovations are perceived to distract the established work practices and routines, organizational norms and procedures, and belief systems of the individuals. Generally, the majority of individuals do not want to be distanced from the current practices because they are well accommodated to routine procedures and practices. Consumers resist innovation because they might fear not receiving the promised value of innovative products due to unfamiliarity with such products and familiarity with the existing products [68]. The level and the extent of resistance on the part of the customers decide the fate of innovations. People fear the changes caused by innovative products and this fear results in resistance to innovation [69]. There are certain functional, financial, economic, usage, emotional and risk, and value barriers that also cause resistance to innovation [70]. There are two types of resistance to innovations. One is active resistance to innovation and the other is passive resistance to innovation. The first one refers to the negative response to innovation resulting from the psychological and functional barriers and the second one mentions to the user's inclination to resist the proposed innovation. Such personal inclination is unconscious and exists even before the innovation [71].

6.2.7 Theory of Planned Behavior (TPB)

The theory of planned behavior refers to the interpretation of personal behavior based on individual thinking. This theory assumes that the determinants of an individual's behavior are based on three main aspects: the someone review of the situations around them, which is called attitude; the influence of the environment, such as inner circle, which is called social norms; and the individual's sense of ability to perform the behavior based on internal capabilities and opportunities [72–74]. Ajzen [75] presented the TPB, which is a strong contribution to the field of social behavior. This theory is effective at both the strategic and organizational levels. It addresses the ambiguity surrounding understanding human behavior. It is based on the causes that result from human behavior. Usually requires understanding the causes of behavior on the one hand and goals on the other [76]. Based on the situations an individual experiences, the process of planning for the future takes place. Therefore, the individual evaluates these situations and justifies their behavioral priorities. So, situations are one of the important components of the theory of planned behavior. Emotional norms recreate a prominent role in producing planned behavior. Subjective norms provide benchmarks for assessing an individual's potential and the challenges of reality. They thus contribute to determining the level and type of planned behavior. Perceived behavioral control can provide signals of the strength or weakness of behavioral performance and is therefore an expression of an individual's self-efficacy [77]. One of the observed studies that demonstrated the effectiveness of the theory of intended behavior is Armitage and Conner [78]. This study presented a meta-analytic review of the theory's role in predicting intentions and future behavior. It revealed the theory's role in varying human behavior. Perceived behavioral control was a factor that contributed 39% to intentions and behaviors. Furthermore, subjective norms appeared as a weaker indicator, indicating that planned behavior is based on external rather than internal influences. Furthermore, other studies have been conducted in several areas, such as exercise, dietary choices, drug use, online consumer behavior, behavior toward climate change, and dietary modification [76, 79–82]. All of these studies have verified the effectiveness of this theory in revealing the nature of human behavior based on events and situations, as well as its role in expressing the habits that drive humans. Hence, it highlights the importance of planned behavior compared to random behavior.

6.2.8 Decomposed Theory of Planned Behavior (DTPB)

This is a conceptual extension of the Theory of Planned Behavior (TPB). Compared to TPB, which focuses on attitude, subjective norm, and behavioral control, DTPB discusses how to delve deeper into these three elements to achieve a more comprehensive understanding of their impact on technology acceptance and adoption [83]. It is therefore based on three cognitive beliefs: utility, convenience, and compatibility. Utility is the perception that an individual will benefit from using the technology. Convenience, on the other hand, implies that there is no risk associated with the new technology, on the contrary, it will enhance the individual's skills. Compatibility refers to the fact that innovations are more compatible with the current needs of individuals. This theory was suggested by Taylor and Todd [84]. They argue that it is derived from the technology acceptance model and the diffusion of innovation theory. Therefore, this theory contributes to shaping attitudes about technology adoption. According to this theory, the social norm is divided into two influences: supervisor influence and peer influence. Supervisor influence means authoritarianism. It expresses the extent of influence exerted by superiors [85]. Peer influence, on the other hand, refers to the influence of colleagues and friends on employees' intentions and behaviors. Thus, influence can be either soft or hard, meaning it is either acquired by the individual or imposed by a higher authority. According to this theory, perceived behavioral control is divided into three components: conditions and resources, technological facilitation, and self-efficacy. Conditions and resources relate to the availability of the technological infrastructure [86]. Self-efficacy describes the skills an employee should possess, which are typically aligned with the technological infrastructure. Technology facilitation refers to the fact that the introduction of technology is not sudden but rather based on pre-planned and detailed plans [83]. The DTPB approach has been adopted in some empirical studies. For example, in electric vehicle usage intentions, Sierzchula et al. (2015) examined the effect of compatibility and interest levels on individuals' attitudes. The study provided insights into electric vehicle usage intentions. In another context, Nyasulu and Chawinga [85] applied to the DTPB to assess the potential for changing intentions toward educational platforms. This study examined individuals' intentions toward using WhatsApp in e-learning. The study found that perceived behavioral control is influenced by resources and efficacy. Moreover, DTPB has been applied in other fields such as games [87].

6.2.9 Technology Acceptance Model (TAM)

Davis [88] developed the TAM model. TAM explains the acceptability of technology by individuals. The basic premise of this model is motivation. Its use is linked to the individual's perceived benefit [89]. The two human objectives of TAM are motivation and advanced use. It determines the extent to which an individual relies on a technology when the outcomes of its facilitators are clear. This model has achieved widespread use across a range of industries. TAM is characterized by its adaptability to a variety of contexts. Doleck et al. [90] fortified this model in an empirical study. The outcome is to identify the extent to which e-learning platforms can be used. They found that acceptance of technology requests is based on learning, organization, and proximity to the technology. Therefore, there is no isolation between them. TAM provides deep insight and understanding of individuals' intentions to rely on digital learning platforms. With the emergence of artificial intelligence technologies, the use of TAM has expanded and has encompassed several fields, including healthcare [91] and e-commerce [92]. Despite its widespread success, Tarhini et al. [93] has raised some criticisms of TAM. After focusing on the invasion of cultural and geographical elements, cultural factors have become more important. These factors address social and cognitive influences. Although the concept of acceptance and its application contribute significantly to our understanding of technology acceptance, its expansion highlights the position of seeing a broader range of issues. Because technology continues to permeate various aspects of work, it is imperative to adapt expanded models to understand the complex interactions of individual and organizational interactions.

Table 6.1 presents a structured interview of nine established theories related to technology adoption, behavioral intentions, and innovation diffusion. Each theory is evaluated based on its founder(s), year of establishment, central concept, strengths, and weaknesses. These theories address multiple fields, including psychology, sociology, marketing, information systems, and communication. The table serves as a summary of these theories, providing the reader with a quick overview. Furthermore, this table allows for comparisons between these theories based on the available criteria, for example, comparison in terms of time, strengths and weaknesses, etc.

Table 6.1 Comparison of key technology adoption and behavioral theories

No	Theory	Founder(s)	Concept	Strength	Weakness
1	Diffusion of innovation theory	[94]	Creation features: comparative benefit, compatibility, intricacy, trialability, observability	Widely applicable across fields; explains stages of adoption	Less effective at predicting individual behavior; assumes linear progression
2	Social support theory	[9]	As belonging to a certain social network and one's belief that he/she is important member of the social network to which he/she belongs	Emphasizes the human/social aspect of adoption and behavior change	Often vague in measuring the actual impact of each support type
3	Social presence theory	[30, 31]	A person's awareness that he is a real person when communicating through social media platforms	Crucial for understanding online and mediated communication	Limited to communication context; doesn't fully address technology adoption
4	Technology acceptance model (TAM)	[95]	Perceived usefulness (PU), perceived ease of use (PEOU)	Simplicity and strong empirical support	Ignores social and contextual factors; limited to workplace systems
5	Innovation resistance theory	[96]	Usage borders, value fences, risk barricades, tradition fences, image barriers	Addresses why people resist innovations, not just why they adopt them	Focuses more on barriers than enablers; sometimes lacks predictive power
6	Theory of planned behavior (TPB)	[76]	Perspective toward conduct, emotional norms, sensed behavioral management	Strong predictive power for intentional behavior	Assumes behavior is planned; neglects emotional/ spontaneous factors
7	Decomposed TPB (DTPB)	[84]	Decomposed into: attitude (PU, PEOU), emotional norms, perceived behavioral control (self-efficacy, resources)	More detailed and specific than TPB	Complexity in measurement; similar constructs can overlap

(continued)

Table 6.1 (continued)

No	Theory	Founder(s)	Concept	Strength	Weakness
8	UTAUT	Venkatesh et al. [34]	Explain user intentions to adopt and use technology	High Predictive Power	Negligence of Expressive and Cultural Issues
9	UTAUT2	[97]	Predicting users' behavioral intentions and actual use of new technologies	Comprehensive; tailored for consumer context	Complexity and redundancy in constructs; context-specific limitations

References

1. Wang, X.: The effect of inconsistent word-of-mouth during the service encounter. J. Serv. Mark. **25**(4), 252–259 (2011)
2. Lee, S., Choeh, J.Y.: Using the social influence of electronic word-of-mouth for predicting product sales: the moderating effect of review or reviewer helpfulness and product type. Sustainability **12**(19), 7918–7952 (2020)
3. Lee, S., Choeh, J.Y.: The impact of online review helpfulness and word of mouth communication on box office performance predictions. Humanit Soc Sci Commun **7**(1), 1–12 (2020)
4. Lee, K., Lee, B., Oh, W.: Thumbs up, sales up? The contingent effect of Facebook likes on sales performance in social commerce. J. Manag. Inf. Syst. **32**(4), 109–143 (2015)
5. Mehta, R.: Understanding perceived retail crowding: a critical review and research agenda. J. Retail. Consum. Serv. **20**(6), 642–649 (2013)
6. Ali Abbasi, G., Abdul Rahim, N.F., Wu, H., Iranmanesh, M., Keong, B.N.C.: Determinants of SME's social media marketing adoption: competitive industry as a moderator. SAGE Open **12**(1), 1–19 (2022)
7. Cohen, S.: Issues in the study and application of social support. Social Support and Health/ Academic (1985)
8. Barrera, M., Jr.: Distinctions between social support concepts, measures, and models. Am. J. Community Psychol. **14**(4), 413–445 (1986)
9. Cobb, S.: Social support as a moderator of life stress. Psychosom. Med. **38**(5), 300–314 (1976)
10. Kim, H.S., Sherman, D.K., Taylor, S.E.: Culture and social support. American Psychologist **63**(6), 518 (2008)
11. Shumaker, S.A., Brownell, A.: Toward a theory of social support: closing conceptual gaps. J. Soc. Issues **40**(4), 11–36 (1984)
12. Hupcey, J.E.: Clarifying the social support theory-research linkage. J. Adv. Nurs. **27**(6), 1231–1241 (1998)
13. Wills, T. A.: Social support and interpersonal relationships. Prosocial Behavior: Review of Personality and Social Psychology, 12 (1991)
14. House, J. S., Kahn, R. L.: Measures and concepts of social support. In: Cohen, S., Syme, S. L. (eds.) Social Support and Health, pp. 83–108. Academic Press, Orlando (1985)
15. Langford, C.P.H., Bowsher, J., Maloney, J.P., Lillis, P.P.: Social support: a conceptual analysis. J. Adv. Nurs. **25**(1), 95–100 (1997)
16. Crocker, J., Canevello, A.: Creating and undermining social support in communal relationships: the role of compassionate and self-image goals. J. Pers. Soc. Psychol. **95**(3), 555 (2008)
17. Lakey, B., Cohen, S.: Social support and theory. Social support measurement and intervention: a guide for health and social scientists **29**, 29–49 (2000)
18. Sarason, B.R., Pierce, G.R., Shearin, E.N., Sarason, I.G., Waltz, J.A., Poppe, L.: Perceived social support and working models of self and actual others. J. Pers. Soc. Psychol. **60**(2), 273 (1991)

19. DeLongis, A., Holtzman, S.: Coping in context: the role of stress, social support, and personality in coping. J. Pers. **73**(6), 1633–1656 (2005)
20. Tam, K.M., Cheung, S.Y.: Measuring physical activity self-efficacy, self-regulation, social support among Hong Kong working adults: a validation study. J. Phys. Educ. **7**(1), 66–73 (2020)
21. Norris, F.H., Kaniasty, K.: Received and perceived social support in times of stress: a test of the social support deterioration deterrence model. J. Pers. Soc. Psychol. **71**(3), 498 (1996)
22. Sarason, B. R., Pierce, G. R., Sarason, I. G.: Social support: the sense of acceptance and the role of relationships (1990)
23. Scarpetta, F.: Practices to display social presence: a study in a shared mediated environment. PsychNology J. **6**(1) (2008)
24. Tu, C.H.: The impacts of text-based CMC on online social presence. J. Interact. Online Learn. **1**(2), 1–24 (2002)
25. Arbaugh, J.B.: Does academic discipline moderate CoI-course outcomes relationships in online MBA courses? The Internet and Higher Education **17**, 16–28 (2013)
26. Eneizan, B., Mohammed, A.G., Alnoor, A., Alabboodi, A.S., Enaizan, O.: Customer acceptance of mobile marketing in Jordan: an extended UTAUT2 model with trust and risk factors. Int. J. Eng. Bus. Manage. **11**(11), 184797901988948 (2019)
27. Kreijns, K., Xu, K., Weidlich, J.: Social presence: conceptualization and measurement. Educ. Psychol. Rev. **34**(1), 139–170 (2022)
28. Remesal, A., Colomina, R.: Social presence and online collaborative small group work: a socioconstructivist account. Comput. Educ. **60**(1), 357–367 (2013)
29. Shin, D.H.: Defining sociability and social presence in social TV. Comput. Hum. Behav. **29**(3), 939–947 (2013)
30. Short, C.R., Kinden, D.A., Stith, R.: Fetal and neonatal development of the microsomal monooxygenase system. Drug Metab. Rev. **5**(1), 1–42 (1976)
31. Short, J., Williams, E., Christie, B.: The Social Psychology of Telecommunications. John Wiley & Sons, London (1976)
32. Oyarzun, B., Barreto, D., Conklin, S.: Instructor social presence effects on learner social presence, achievement, and satisfaction. TechTrends **62**(6), 625–634 (2018)
33. Richardson, J., Swan, K., Lowenthal, P., Ice, P.: Social presence in online learning: past, present, and future. In Global Learn, pp. 477–483. Association for the Advancement of Computing in Education (AACE) (2016)
34. Venkatesh, V., Morris, M.G., Davis, G.B., Davis, F.D.: User acceptance of information technology: toward a unified view. MIS Q. **27**(3), 425–478 (2003)
35. Dwivedi, Y.K., Rana, N.P., Jeyaraj, A., Clement, M., Williams, M.D.: Re-examining the unified theory of acceptance and use of technology (UTAUT): towards a revised theoretical model. Inf. Syst. Front. **21**, 719–734 (2019)
36. Mensah, I.K., Khan, M.K.: Unified theory of acceptance and use of technology (UTAUT) model: factors influencing mobile banking services' adoption in China. SAGE Open **14**(1), 21582440241234230 (2024)
37. Alshehri, M., Drew, S., Alhussain, T., Alghamdi, R.: The effects of website quality on the adoption of E-government service: an empirical study applying UTAUT model using SEM (2012). arXiv:1211.2410
38. Alowayr, A.: Determinants of mobile learning adoption: extending the unified theory of acceptance and use of technology (UTAUT). Int J Inf Learn Technol **39**(1), 1–12 (2022)
39. Efiloğlu Kurt, Ö., Tingöy, Ö.: The acceptance and use of a virtual learning environment in higher education: an empirical study in Turkey, and the UK. Int. J. Educ. Technol. High. Educ. **14**, 1–15 (2017)
40. Venkatesh, V. (2003). Technology acceptance model and the unified theory of acceptance and use of technology: A review and directions for future research.
41. Chang, A.: UTAUT and UTAUT 2: a review and agenda for future research. The Winners **13**(2), 10–114 (2012)

42. Raman, A., Don, Y.: Preservice teachers' acceptance of learning management software: an application of the UTAUT2 model. Int. Educ. Stud. **6**(7), 157–164 (2013)
43. Celik, H.:. Customer online shopping anxiety within the unified theory of acceptance and use technology (UTAUT) framework. Asia Pacific J. Market. Logist. **28**(2) (2016)
44. Workman, M.: New media and the changing face of information technology use: the importance of task pursuit, social influence, and experience. Comput. Hum. Behav. **31**, 111–117 (2014)
45. Oh, J.C., Yoon, S.J.: Predicting the use of online information services based on a modified UTAUT model. Behav. Inf. Technol. **33**(7), 716–729 (2014)
46. Abd Aziz, N.N., Samad, S.: Innovation and competitive advantage: moderating effects of firm age in foods manufacturing SMEs in Malaysia. Proc. Econom. Finance **35**, 256–266 (2016)
47. Jabbar, M.N., Mahmood, W., Qambar, G.: Mediating role of organizational commitment and work environment on the relationship between transformational leadership and job satisfaction. J. Talent Dev. Excellence **12**(2), 3974–3988 (2020)
48. Dingfelder, H.E., Mandell, D.S.: Bridging the research-to-practice gap in autism intervention: an application of diffusion of innovation theory. J. Autism Dev. Disord. **41**, 597–609 (2011)
49. Greenhalgh, T., Robert, G., Macfarlane, F., Bate, P., Kyriakidou, O.: Diffusion of innovations in service organizations: systematic review and recommendations. Milbank Q. **82**(4), 581–629 (2004)
50. Hamid, S. A. K., Bashir, H., Haridy, S., & Shamsuzzaman, M.: Factors affecting innovation in the service industry: a literature-based model. In: 2022 Advances in Science and Engineering Technology International Conferences (ASET), pp. 1–5. IEEE (2022)
51. Min, S., Kim, J., Sawng, Y.W.: The effect of innovation network size and public R&D investment on regional innovation efficiency. Technol. Forecast. Soc. Chang. **155**, 119998 (2020)
52. Simpson, D.D.: A conceptual framework for transferring research to practice. J. Subst. Abuse Treat. **22**(4), 171–182 (2002)
53. Miller, R. L.: Rogers' innovation diffusion theory (1962, 1995). In: Information Seeking Behavior and Technology Adoption: Theories and Trends, pp. 261–274. IGI Global (2015)
54. Orr, G.: Diffusion of innovations, by Everett Rogers (1995). Retrieved 21 Jan 2005 (2003)
55. Rogers, E.M.: Diffusion of Innovations, 5th edn. Free Press, New York (2003)
56. Scott, S., McGuire, J.: Using diffusion of innovation theory to promote universally designed college instruction. Int. J. Teach. Learn. Higher Educ. **29**(1), 119–128 (2017)
57. Abbas, A., Luo, X., Shahzad, F., Wattoo, M.U.: Optimizing organizational performance in manufacturing: the role of IT capability, green supply chain integration, and green innovation. J. Clean. Prod. **423**, 138848 (2023)
58. Owolabi Yusuf, M.B., Mat Derus, A.: Measurement model of corporate zakat collection in Malaysia: a test of diffusion of innovation theory. Humanomics **29**(1), 61–74 (2013)
59. Sartipi, F.: Diffusion of innovation theory in the realm of environmental construction. J. Constr. Mater. **1**(4), 2–4 (2020)
60. Murray, C.E.: Diffusion of innovation theory: a bridge for the research-practice gap in counseling. J. Couns. Dev. **87**(1), 108–116 (2009)
61. Al-Abrrow, H., Alnoor, A., Abbas, S.J.O.M.J.: The effect of organizational resilience and CEO's narcissism on project success: organizational risk as a mediating variable, **16**(1), 1–13 (2019)
62. Hietschold, N., Reinhardt, R., Gurtner, S.: Who put the "NO" in Innovation? Innovation resistance leaders' behaviors and self-identities. Technol. Forecast. Soc. Chang. **158**, 120177 (2020)
63. Mellor, A., Mobilia, M., Redner, S., Rucklidge, A.M., Ward, J.A.: Influence of Luddism on innovation diffusion (2015). arXiv:1505.02020
64. Talwar, S., Talwar, M., Kaur, P., Dhir, A.: Consumers' resistance to digital innovations: a systematic review and framework development. Australian Market. J. **28**(4), 286–299 (2020)
65. Sadiq, M., Le-Dinh, T., Kien Tran, T., Chien, F., Hien Phan, T. T., Quang Huy, P.: The role of green finance, eco-innovation, and creativity in the sustainable development goals of ASEAN countries. Economic Research-Ekonomska istraživanja **36**(2) (2023)

66. Szmigin, I., Foxall, G.: Three forms of innovation resistance: the case of retail payment methods. Technovation **18**(6–7), 459–468 (1998)
67. Mohd Ishak, S.S., Newton, S.: An innovation resistance factor model. Constr. Econ. Build. **16**(3), 87–103 (2016)
68. Ma, L., Lee, C.S.: Understanding the barriers to the use of MOOCs in a developing country: an innovation resistance perspective. J Educ Comput Res **57**(3), 571–590 (2019)
69. Kaur, N., Kaur, P.: What drives innovation in micro, small, and medium enterprises? J. Public Aff. **21**(2), e2336 (2021)
70. Fan, F., Lian, H., Wang, S.: Can regional collaborative innovation improve innovation efficiency? An empirical study of Chinese cities. Growth Chang. **51**(1), 440–463 (2020)
71. Heidenreich, S., Kraemer, T.: Innovations—doomed to fail? Investigating strategies to overcome passive innovation resistance. J. Prod. Innov. Manag. **33**(3), 277–297 (2016)
72. Ajzen, I.: The theory of planned behavior: frequently asked questions. Human Behavior Emerg. Technol. **2**(4), 314–324 (2020)
73. Mandal, A., Johansson, C., Lindelöw, D.: Exploring walking from the perspective of theory of planned behavior. Transp Res Interdisc. Perspect. **22**, 100931 (2023)
74. Van der Velde, L.A., van Dijk, W.W., Numans, M.E., Kiefte-de Jong, J.C.: Extending the theory of planned behavior for explaining dietary quality: the role of financial scarcity and food insecurity status. J. Nutr. Educ. Behav. **54**(7), 636–646 (2022)
75. Ajzen, I.: From intentions to actions: A theory of planned behavior. In: Kuhl, J., Beckmann, J. (eds.) Action Control: From Cognition to Behavior, pp. 11–39. Springer, Berlin (1985)
76. Ajzen, I.: The theory of planned behavior. Organ. Behav. Hum. Decis. Process. **50**(2), 179–211 (1991)
77. Terry, D.J., O'Leary, J.E.: The theory of planned behaviour: the effects of perceived behavioural control and self-efficacy. Br. J. Soc. Psychol. **34**(2), 199–220 (1995)
78. Armitage, C. J., Conner, M.: Efficacy of the theory of planned behaviour: A meta-analytic review. Br. J. Soc. Psychol. **40**(4), 471–499 (2001)
79. Conner, M., Armitage, C.J.: Extending the theory of planned behavior: a review and avenues for further research. J. Appl. Soc. Psychol. **28**(15), 1429–1464 (1998)
80. George, J.F.: The theory of planned behavior and Internet purchasing. Internet Res. **14**(3), 198–212 (2004)
81. McDermott, M.S., Oliver, M., Svenson, A., Simnadis, T.: A systematic literature review and meta-analysis: the Theory of Planned Behavior's application to understand and predict nutrition-related behaviors in youth. Appetite **85**, 336–346 (2015)
82. Norman, P., Conner, M.: The theory of planned behaviour and binge drinking among undergraduate students: assessing the impact of habit strength. Addict. Behav. **31**(4), 758–770 (2006)
83. Moons, I., De Pelsmacker, P.: An extended decomposed theory of planned behaviour to predict the usage intention of the electric car: a multi-group comparison. Sustainability **7**(5), 6212–6245 (2015)
84. Taylor, S., Todd, P.A.: Understanding information technology usage: a test of competing models. Inf. Syst. Res. **6**(2), 144–176 (1995)
85. Nyasulu, C., Dominic Chawinga, W.: Using the decomposed theory of planned behaviour to understand university students' adoption of WhatsApp in learning. E-Learning and Digital Media **16**(5), 413–429 (2019)
86. Ghani, W.A.W.A.K., Rusli, I.F., Biak, D.R.A., Idris, A.: An application of the theory of planned behaviour to study the influencing factors of participation in source separation of food waste. Waste Manage. **33**(5), 1276–1281 (2013)
87. Huang, C.-M., Huang, S.-F., Chen, Y.-T., Chang, C.-H., Chien, H.-C., Chang, Y.-J., Huang, K.-Y., Guo, J.-L.: Applicability of the decomposed theory of planned behavior for the evaluation of community-dwelling older adults' acceptance in continuous usage of robot-assisted board games for cognitive training. Digital Health **11**, 20552076241312576 (2025)
88. Davis, F. D.: Perceived usefulness, perceived ease of use, and user acceptance of information technology. MIS Quarter. **13**(3), 319–340 (1989)

89. Silva, P.: Davis' technology acceptance model (TAM) (1989). Information seeking behavior and technology adoption: theories and trends 205–219 (2015)
90. Doleck, T., Bazelais, P., Lemay, D.J.: Is a general extended technology acceptance model for e-learning generalizable? Knowl. Manage. E-Learn. **10**(2), 133–147 (2018)
91. Holden, R.J., Karsh, B.T.: The technology acceptance model: its past and its future in health care. J. Biomed. Inform. **43**(1), 159–172 (2010)
92. Nagy, S., Hajdu, N.: Consumer acceptance of the use of artificial intelligence in online shopping: evidence from Hungary (2023). arXiv:2301.01277
93. Tarhini, A., Hone, K., Liu, X.: The effects of cultural dimensions and demographic characteristics on e-learning acceptance. Comput. Hum. Behav. **41**, 153–163 (2016)
94. Rogers, E.M.: Diffusion of Innovations. Free Press (1962)
95. Davis, F. D.: A technology acceptance model for empirically testing new end-user information systems. Doctoral dissertation, MIT Sloan School of Management (1986)
96. Ram, S., Sheth, J.N.: Consumer resistance to innovations: the marketing problem and its solutions. J. Consum. Mark. **6**(2), 5–14 (1989)
97. Venkatesh, V., Thong, J.Y.L., Xu, X.: Consumer acceptance and use of information technology: extending the unified theory of acceptance and use of technology. MIS Q. **36**(1), 157–178 (2012)

Part III
ANN and FsQCA Applications

Chapter 7
Hybrid SEM and ANN Approach

Abstract Exploring the integration of SEM and ANN for uncovering complex and non-linear relationships in social science research. The chapter begins with introducing the foundation of ANN. It then reviews the application of ANN across diverse academic domains. The chapter provides a practical guide for implementing ANN. Researchers are empowered to combine the causal inference capabilities of SEM with the high predictive accuracy of ANN to capture linear and nonlinear links. Ultimately, offering a more robust and insightful understanding of empirical data.

Learning Objectives

1. This chapter aims to discuss the concept of ANN.
2. This chapter discusses the uses of ANN in previous literature.
3. The chapter explains in detail the mechanism of implementing ANN step by step.

7.1 Introduction

The theoretical field of industry applications concentrates on comprehending the interactions between antecedents and outcomes. Much of the previous literature has relied on applying linear methods to discover lined and fundamental relations among variables [1, 2]. Common statistical methods include two types: first-generation and second-generation tools. First-generation tools include regression analysis, confirmatory factor analysis, and exploratory analysis. Second-generation tools include structural equation modeling [3]. The increasing complexity of phenomena in the social sciences has led to insufficient reliance on second-generation tools. In addition, second-generation tools help capture linear relationships but fail to explain nonlinear relationships or provide configurations [4]. To overcome the limitations of linear methods, thinkers have employed numerous methodological methods to investigate the interplay between causes and outcomes and to go beyond linear relationships

X. Chew et al., *Partial Least Squares Structural Equation Modeling and Complementary Methods in Business Research*, Information Systems Engineering and Management 67, https://doi.org/10.1007/978-3-032-01055-1_7

[5]. Scientists have recommended using ANN to classify non-compensatory relations among factors. ANN investigation provides more conclusive evidence about the most important determinants of output. The literature confirms that using the ANN method does not require a large sample size [6]. Furthermore, several steps must be followed to implement and employ the ANN method.

SPSS software is used to employ ANN. This investigation completes the work of PLS-SEM by verifying the validity of PLS-SEM results and presenting nonlinear relationships. To this end, 90% of the data should be selected for exercise and 10% for testing. This approach helps reduce the problem of overfitting by ensuring that all data were tested in the training and testing processes and that there is no bias [7]. There are many activation functions, but previous literature recommends using the sigmoid function because it provides more accurate results, improves prediction accuracy, and reduces errors. To determine prediction accuracy, the root mean square deviation (RMSD) is employed to measure the prediction correctness of the neuronal model [8]. The ANN technique resembles the mortal brain and includes cells and units that keep knowledge and make it available for use. A multi-layer feed-forward approach should be implemented, consisting of input, processing, and output layers [9]. Each layer consists of neurons that contribute to feeding the model from infusion to result via the invisible layer. Each cell in the ANN model calculates outputs based on the amount of energy, and the weights are signified by the contribution constituent (i), the concealed layer (j), and the result layer (k), as shown in the following equations [10].

$$net_j^h = \sum_{i=1}^{N} Wjix_i \text{ and } Y_i = f\left(net_j^h\right) \tag{7.1}$$

For the k-th output neuron,

$$net_k^0 = \sum_{j=1}^{J+1} V_{kj}y_j \text{ and } o_k = f\left(net_k^0\right). \tag{7.2}$$

$$f(net) = \frac{1}{1 + e^{-\lambda net}} \tag{7.3}$$

$$V_{kj}(t+1) = v_{kj}(t) + c\lambda(d_k - o_k)o_k(1 - o_k)y_i(t) \tag{7.4}$$

$$W_{ji}(t+1) = W_{ji}(t) + c\lambda^2 y_j\left(1 - y_j\right)x_i(t)\left(\sum_{k=1}^{k}(d_k - o_k)o_k(1 - o_k)V_{kj}\right) \tag{7.5}$$

$$SSE = \frac{1}{2P}\sum_{p=1}^{p}\sum_{k=1}^{k}(d_{pk} - o_{pk})^2 \tag{7.6}$$

ANN analysis captures nonlinear relationships and acts as a black box. The outputs supported by the PLS-SEM investigation serve as inputs to the ANN analysis. ANN results help identify the factors that contribute most to the outcome studied. Furthermore, results that capture seemingly complex nonlinear relationships explain and complement the work of second-generation statistical approaches. ANNs are particularly effective in interpreting complex models with multiple paths [3].

7.2 Artificial Neural Networks Based on Practical Application

In this section, we attempt to explain simply and accurately some applications of artificial neural networks (ANNs). Table 7.1 shown two practical subcategories: artificial neural networks and their application methods. These two subcategories branch into other subcategories. Therefore, it is necessary to clarify all the details related to this important technique. In the social sciences, the reliability of a study's scale is usually analyzed. Called the assessment of a measurement model. This type of analysis relies on two statistical tools: discriminant validity and convert validity. The first includes three indicators: cross-loading, fornell and lacoller, and heterotrait-monotrait. The second also includes three indicators: average variance extracted, composite reliability, and factor loading. If one or more (depending on the indicators the researcher uses) match the standard values, the scale is considered reliable. In light of these results, the researcher moves on to the following examination: the assessment of a structural model, which is used to analyse the study's hypotheses. After conducting hypothesis testing and obtaining results, researchers turn to ANN testing to confirm hidden patterns in the study model. This is done by answering the question about the validity of the results revealed in the hypothesis test. This is done by testing nonlinear relationships. In this analysis process, researchers use a method called tenfold neural networks. This method divides the data into ten layers, indicating that some nodes within the network are being neglected. Consequently, results appear that are contrary to reality or not close to reality. In other words, the importance level of some elements may not reflect their natural state. This affects the model's goal of reducing prediction bias. Table 7.1 includes some previous studies that neglected some processes.

Table 7.1 shows that the central issue depicted in the previous literature is the presence of various settings when using the ANN and contribution processes. Table 7.1 shows that 45% of studies used a single hidden layer, while only 8% used two invisible layers. Although the activation procedure includes two kinds, the stress was on the sigmoid process. 52% of investigations used this procedure. Additionally, the studies used a tenfold technique to display the function [2, 12–20, 36–50], Sharma et al. [51].

Table 7.1 ANN process

References	Automatic building selection	Custom building							
		Concealed layers		Activation purpose		Production layer			
		One	Two	Hyperbolic tangent	Sigmoid	Identity	Softmax	Hyperbolic tangent	Sigmoid
[11]			√						
[12–20]	√								
[21–24]			√		√				√
[25, 26]		√			√				
[27–30], Leong et al. [31, 32]		√			√				
[33–35]		√			√				√

Regarding the contribution process, the problem arises in the technique used to calculate the contribution. Several methods are available for calculating the contribution, but the most studies rely on only two. Sensitivity analysis and partial derivatives are the most widely used methods, indicating the neglect of other methods [24, 27–30, 46, 52–59], Leong et al. [31, 32].

7.3 How to Conduct ANN Analysis in a Regular and Even Manner Guide

7.3.1 Overview of ANN Analysis

Due to the limitations of SEM, an integrated approach SEM-ANN is being used to address the shortcoming of SEM methods. ANNs help to determine and understand the no-compensatory and no-linear relationship among the study factors. The SEM-ANN approach is a sophisticated method that is also used to authenticate the consequences obtained through SEM and measure the connection between the predictors and output variables [24]. ANNs method also allows to rank and authenticate the major determinants in a proposed research model. ANN can be considered to function like human brain as it is based on the store the relevant data and process the data in sequential way and combine some complex statistical techniques to generate the results. ANN approach is used across various fields of study including e-commerce, aviation, manufacturing, engineering, medical and mobile payment methods. Most studies using ANNs depend on using multilayer perceptron with a 'feed forward and back propagation' algorithm [52]. In this method, important determining variables from the PLS path investigation are utilized as infusion neurons. Along with this, a sigmoid operation is used to trigger the outcome and the masked layer. In this study, a neuronal net algorithm was applied while using a tenfold ANN examination to overcome the potential case of over-suitable with tenner percent of the information used for challenging and the residual information applied for exercise events. The mean root means square deviation (RMSD). (RMSD) was utilized to degree the prognostic correctness of the ANN method. The golly-of-fit index of the ANN model can be intended by using the ($R^2 = 1 - \text{RMSE}/(\text{Sy}^\wedge 2)$. That is alike to R^2 in SEM investigation [6]. Figure 7.1 shown ANN process.

It is mandatory that when using ANN models then input for neuron must be major factors deduced from PLS-SEM method. But when to determine the mediation then all the paths must be accepted and if not accepted then mediation and interactive variables need not to be included in ANN analysis. Under ANN analysis, predictors, mediators, and interactive variables are taken as covariates and control variables are taken as factors. There are two major advantages of employing a multi-analytical two steps SEM-ANN system. One helps to validate the findings from SEM and second it also capture active non-linear relationships between predictors and outcome

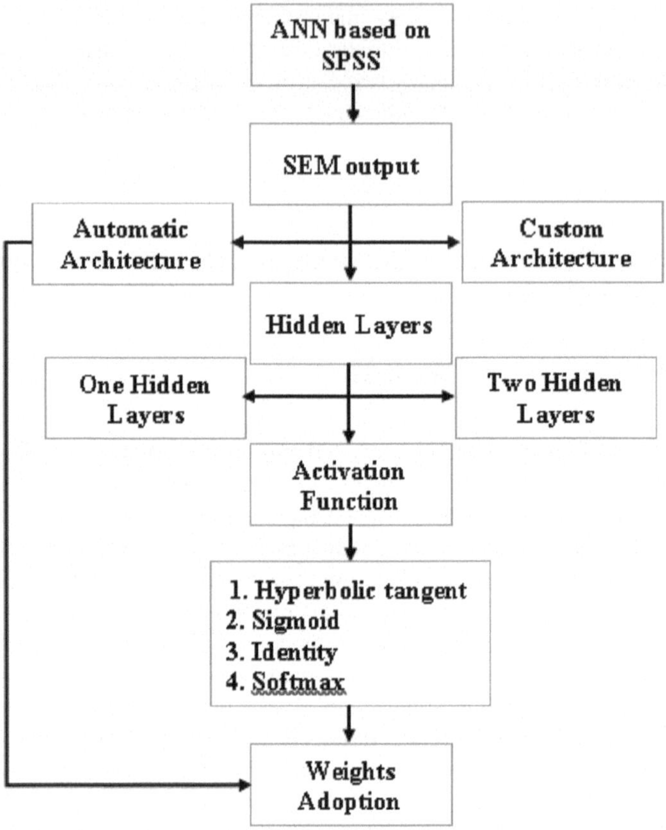

Fig. 7.1 ANN process

variables and give additional correct value of each predictor's respective power in the relationship between predictors and outcome variables.

7.3.2 Performing Artificial Neural Network Analysis

There are four widely used scales for measuring variables: nominal, ordinal, interval, and ratio. The nominal scale is utilized for qualitative factors such as gender, civil ID number, and university ID number. Numbers are used instead of labels; for example, 1 represents males and 2 represents females. These numbers represent two distinct groups. The ordinal scale is higher than the nominal scale and classifies variables in ascending or descending order, such as educational qualifications. In the interval scale, numbers carry a quantitative meaning and contain units of measurement, such as the difference between grades or temperature. Finally, the ratio scale is suitable

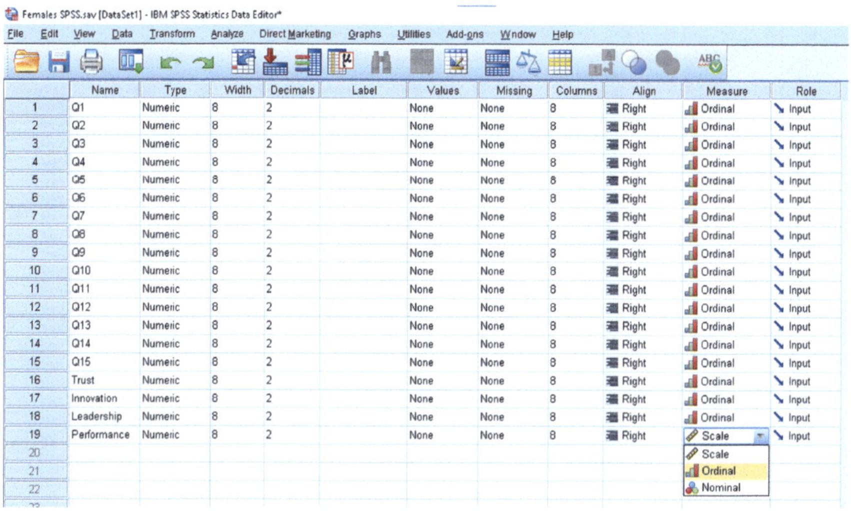

Fig. 7.2 Data process

for physical variables such as weight, speed, height, and others [60]. Figure 7.2 illustrates variables scales. The first step involves transforming the data into ordinal values for the dependent variable.

As previously mentioned, the advantage of applying the ANN approach lies in the robustness of the results of linear methods such as regression and in capturing nonlinear relationships. Many tools and programs exist for implementing the nonlinear approach. However, SPSS is the most popular program for implementing the ANN approach. Neural networks can be tested from the analysis menu and initiated using such a methodology, as illustrated in Fig. 7.3.

Earlier studies recommend that the percentage of the training data should be 90%. In terms of the testing data, the percentage must be 10%. A data distribution methodology is used for training and testing to avoid overfitting. This methodology helps avoid bias and deliver more accurate results. Hence, the ANN approach supplies the grade and validates the influential features of a model. ANN is a companion of the mortal brain. It remembers understanding through continuous exercise procedures and stores obtained knowledge as synaptic importance. The contribution of the neuron to the ANN method should be an important forecaster from the PLS-SEM investigation. However, in the case of the mediate and interactive variables, all paths must be accepted. Otherwise, the interactive or mediating variable should be excluded from the ANN analysis. As depicted in Fig. 7.4, the main, mediating, and moderating variables are treated as covariates. While control variables are treated as factors.

The ANN method collars not just linear but also engaged nonlinear relationships between prototypes and dependent factors, and a more exact measure of each predictor's comparative power as well [6]. Literature has used automatic architecture

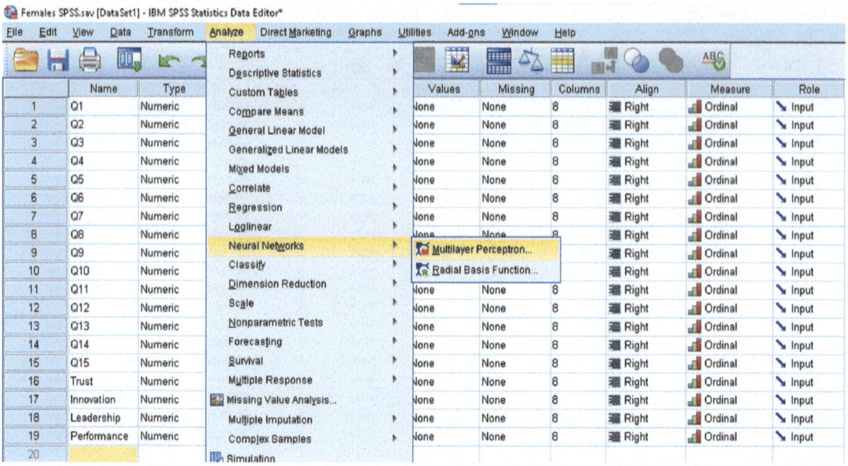

Fig. 7.3 ANN analysis

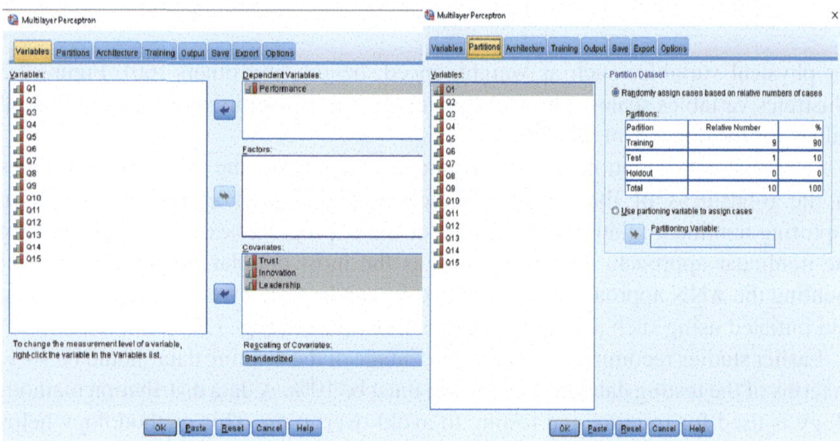

Fig. 7.4 Variables process

selection and custom architecture as shown in Table 7.1 and Fig. 7.5. Hence, we may cite and justify the selections process based on the table that we mentioned previously about ANN process. The prognostic correctness of the ANN method was slow using the root mean square deviation (RMSD). As indicated, the little and similar RMSD mean standards indicate that the ANN M for the perfect model can suggest increased forecast precision.

A sensitivity analysis assessed each variable's contribution to the performance development. Normalized extent was estimated according to the percentage, leaning on the comparative volume of per infusion neuron area, to give the most lavish comparative importance for the performance model. A goodness-of-fit index was

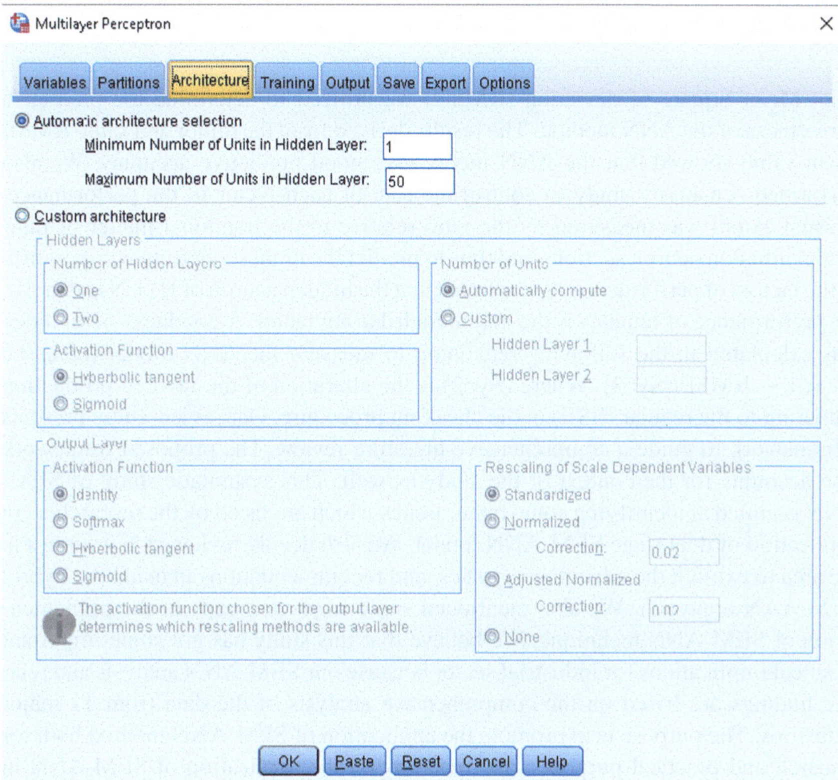

Fig. 7.5 Architecture selection

grasped in the following equation to define the ANN model prognoses: $R^2 = 1 - RMSE/Sy^2$. where (Sy^2) is the contention of the choice creation according to the norm in the testing method. If the data is smallish, complex, and has fewer proportions or features, then neural networks with 1–2 hidden layers would work. When the data has large dimensions or components, then to get an optimum resolution, 3–5 hidden layers can be utilized. A hidden layer in an ANN is a layer between input layers and output layers, where artificial neurons take in a bunch of weighted inputs and produce an output via an activation function. The altitude of the unseeable layer is generally between the sizes of the input and output. It should be 2/3 the size of the input layer plus the size of the output layer. The number of hidden neurons should be less than twice the size of the input layer [61].

7.4 Conclusion of ANN Results

Root Mean Square of deviation (RMSD) we utilized to determine the predictive correctness of the ANN method. The results derived from the minor and same RMSE mean value showed that the ANN model give good predictive accuracy. We also performed sensitivity study to control the role of each factor in the performance. Normal extent was measured by the ratio relative to the fractional matter of individual infusion neuron sections and thus to predict the highest respective importance of the factors of performance. It was found that the hidden neuron of H (1:N) related to the performance of females is the major contributing factor. A goodness-of-fit index was calculated in the following reckoning to measure the ANN model forecasts. $R^2 = 1 - \text{RMSE}/(\text{Sy}^\wedge 2)$. Where (Sy^2) is the alteration of the favored production rendering to the regular (SSE) in the checking procedure. Our review study presents a framework to guide a comprehensive literature review. The proposed framework also accounts for the context of the study as well. This systematic study of SEM-ANN is aimed at identifying some major issues which are faced by the researchers in application of dual stage SEM-ANN model. We also deeply reviewed key studies in the area to explicit the advantages, issues, and recommendations in using the hybrid SEM-AAN approach. We also mentioned some important gaps in the implementation of SEM-ANN technique. We believe that this study has got some important practical implications for industrial sector because our SEM-ANN analysis and relative findings are based on the comprehensive analysis of the data from 11 major industries. The purpose is to promote the application of SEM-ANN method both for research and practical purposes. We recommend the application of SEM-ANN in the technology sector as compared to other industries like small and medium scale construction firms. We also hope that our study will facilitate the future researchers and practitioners by providing an insight details on the use of SEM-ANN method.

References

1. Albahri, A.S., Alnoor, A., Zaidan, A.A., Albahri, O.S., Hameed, H., Zaidan, B.B., Yass, A.A.: Hybrid artificial neural network and structural equation modelling techniques: a survey. Complex Intell. Syst. **8**(2), 1781–1801 (2022)
2. Alnoor, A., Tiberius, V., Atiyah, A.G., Khaw, K.W., Yin, T.S., Chew, X., Abbas, S.: How positive and negative electronic word of mouth (eWOM) affects customers' intention to use social commerce? A dual-stage multi group-SEM and ANN analysis. Int. J. Hum.-Comput. Interact. **40**(3), 808–837 (2024)
3. Ferasso, M., Alnoor, A.: Artificial neural network and structural equation modeling in the future. In: Artificial Neural Networks and Structural Equation Modeling: Marketing and Consumer Research Applications, pp. 327–341. Springer Nature Singapore, Singapore (2022)
4. Alnoor, A., Abbas, S., Khaw, K.W., Muhsen, Y.R., Chew, X.: Unveiling the optimal configuration of impulsive buying behavior using fuzzy set qualitative comparative analysis and multi-criteria decision approach. J. Retail. Consum. Serv. **81**(11), 1–17 (2024)

5. Basco, R., Hair, J.F., Jr., Ringle, C.M., Sarstedt, M.: Advancing family business research through modeling nonlinear relationships: comparing PLS-SEM and multiple regression. J. Fam. Bus. Strat. **13**(3), 1–16 (2022)
6. Khaw, K.W., Alnoor, A., Al-Abrrow, H., Chew, X., Sadaa, A.M., Abbas, S., Khattak, Z.Z.: Modelling and evaluating trust in mobile commerce: a hybrid three stage Fuzzy Delphi, structural equation modeling, and neural network approach. Int. J. Hum.-Comput. Interact. **38**(16), 1529–1545 (2022)
7. Mkedder, N., Bakır, M.: A hybrid analysis of consumer preference for domestic products: combining PLS-SEM and ANN approaches. J. Glob. Mark. **36**(5), 372–395 (2023)
8. Hayat, N., Salameh, A.A., Mamun, A.A., Helmi Ali, M., Makhbul, Z.K.M.: Tax compliance behavior among Malaysian taxpayers: a dual-stage PLS-SEM and ANN analysis. SAGE Open **12**(3), 1–14 (2022)
9. Yin, Y.C., Ahmed, J., Nee, A.Y.H., Hoe, O.K.: The rural consumer adoption of sustainable energy: a PLS-SEM-ANN approach of conceptual model development and cross-country validation of Pakistan and Malaysia. Environ. Sci. Pollut. Res. **30**(3), 5881–5902 (2023)
10. Khaw, K. W., Camilleri, M., Tiberius, V., Alnoor, A., Zaidan, A. S.: Benchmarking electric power companies' sustainability and circular economy behaviors: using a hybrid PLS-SEM and MCDM approach. Environ. Develop. Sustain. **26**(3), 6561–6599 (2024)
11. Abubakar, A.M., Namin, B.H., Harazneh, I., Arasli, H., Tunç, T.: Does gender moderates the relationship between favoritism/nepotism, supervisor incivility, cynicism and workplace withdrawal: a neural network and SEM approach. Tourism Manage. Perspect. **23**, 129–139 (2017)
12. Ahani, A., Rahim, N.Z.A., Nilashi, M.: Forecasting social CRM adoption in SMEs: a combined SEM-neural network method. Comput. Hum. Behav. **75**, 560–578 (2017)
13. Alam, M.Z., Hu, W., Kaium, M.A., Hoque, M.R., Alam, M.M.D.: Understanding the determinants of mHealth apps adoption in Bangladesh: a SEM-Neural network approach. Technol. Soc. **61**, 101255 (2020)
14. Dadashova, B., Arenas-Ramírez, B., Mira-McWilliams, J., Aparicio-Izquierdo, F.: Methodological development for selection of significant predictors explaining fatal road accidents. Accid. Anal. Prev. **90**, 82–94 (2016)
15. Foo, P.Y., Lee, V.H., Tan, G.W.H., Ooi, K.B.: A gateway to realising sustainability performance via green supply chain management practices: a PLS–ANN approach. Expert Syst. Appl. **107**, 1–14 (2018)
16. Hidayat-ur-Rehman, I., Alzahrani, S., Rehman, M.Z., Akhter, F.: Determining the factors of m-wallets adoption. A twofold SEM-ANN approach. PLoS ONE **17**(1), e0262954 (2022)
17. Islam, A.N., Laato, S., Talukder, S., Sutinen, E.: Misinformation sharing and social media fatigue during COVID-19: an affordance and cognitive load perspective. Technol. Forecast. Soc. Chang. **159**, 120201 (2020)
18. Khayer, A., Talukder, M.S., Bao, Y., Hossain, M.N.: Cloud computing adoption and its impact on SMEs' performance for cloud supported operations: a dual-stage analytical approach. Technol. Soc. **60**, 101225 (2020)
19. Kruk, M., Paturej, E., Artiemjew, P.: From explanatory to predictive network modeling of relationships among ecological indicators in the shallow temperate lagoon. Ecol. Ind. **117**, 106637 (2020)
20. Larbi-Siaw, O., Xuhua, H., Owusu, E., Owusu-Agyeman, A., Fulgence, B.E., Frimpong, S.A.: Eco-innovation, sustainable business performance and market turbulence moderation in emerging economies. Technol. Soc. **68**, 101899 (2022)
21. Chen, H., Liu, H., Chu, X., Zhang, L., Yan, B.: A two-phased SEM-neural network approach for consumer preference analysis. Adv. Eng. Inform. **46**, 101156 (2020)
22. Nair, D.J., Rashidi, T.H., Dixit, V.V.: Estimating surplus food supply for food rescue and delivery operations. Socioecon. Plann. Sci. **57**, 73–83 (2017)
23. Raut, R.D., Mangla, S.K., Narwane, V.S., Gardas, B.B., Priyadarshinee, P., Narkhede, B.E.: Linking big data analytics and operational sustainability practices for sustainable business management. J. Clean. Prod. **224**, 10–24 (2019)

24. Raut, R.D., Priyadarshinee, P., Gardas, B.B., Jha, M.K.: Analyzing the factors influencing cloud computing adoption using three stage hybrid SEM-ANN-ISM (SEANIS) approach. Technol. Forecast. Soc. Chang. **134**, 98–123 (2018)
25. Kalinic, Z., Marinkovic, V., Molinillo, S., Liébana-Cabanillas, F.: A multi-analytical approach to peer-to-peer mobile payment acceptance prediction. J. Retail. Consum. Serv. **49**, 143–153 (2019)
26. Liébana-Cabanillas, F., Marinkovic, V., de Luna, I.R., Kalinic, Z.: Predicting the determinants of mobile payment acceptance: a hybrid SEM-neural network approach. Technol. Forecast. Soc. Change **129**, 117–130 (2018)
27. Alkawsi, G.A., et al.: A hybrid SEM-neural network method for identifying acceptance factors of the smart meters in Malaysia: challenges perspective. Alex. Eng. J. **60**(1), 227–240 (2021)
28. Hew, J.J., Badaruddin, M.N.B.A., Moorthy, M.K.: Crafting a smartphone repurchase decision making process: do brand attachment and gender matter? Telematics Inform. **34**(4), 34–56 (2017)
29. Hew, J.J., Leong, L.Y., Tan, G.W.H., Lee, V.H., Ooi, K.B.: Mobile social tourism shopping: a dual-stage analysis of a multi-mediation model. Tour. Manage. **66**, 121–139 (2018)
30. Lee, V.H., Foo, A.T.L., Leong, L.Y., Ooi, K.B.: Can competitive advantage be achieved through knowledge management? A case study on SMEs. Expert Syst. Appl. **65**, 136–151 (2016)
31. Leong, L.Y., Hew, T.S., Ooi, K.B., Chong, A.Y.L.: Predicting the antecedents of trust in social commerce—a hybrid structural equation modeling with neural network approach. J. Bus. Res. **110**(1), 24–40 (2020a). https://doi.org/10.1016/j.jbusres.2019.11.056
32. Leong, L.Y., Hew, T.S., Ooi, K.B., Wei, J.: Predicting mobile wallet resistance: A two-staged structural equation modeling-artificial neural network approach. Int. J. Info. Manage. **51**, 102047 (2020b)
33. Binsawad, M.H.: Corporate social responsibility in higher education: a PLS-SEM neural network approach. IEEE Access **8**, 29125–29131 (2020)
34. Najmi, A., Kanapathy, K., Aziz, A.A.: Exploring consumer participation in environment management: findings from two-staged structural equation modelling-artificial neural network approach. Corp. Soc. Responsib. Environ. Manag. **28**(1), 184–195 (2021)
35. Sternad Zabukovšek, S., Kalinic, Z., Bobek, S., Tominc, P.: SEM–ANN based research of factors' impact on extended use of ERP systems. CEJOR **27**, 703–735 (2019)
36. Almarzouqi, A., Aburayya, A., Salloum, S.A.: Determinants of intention to use medical smartwatch-based dual-stage SEM-ANN analysis. Inform. Med. Unlocked **28**, 100859 (2022)
37. Alnoor, A., Atiyah, A.G.: An SEM-ANN analysis of the influence of strategic merger on strategic postchange of family businesses. Nankai Bus. Rev. Int. (ahead-of-print) (2024)
38. Alnoor, A., Al-Abrrow, H., Al Halbusi, H., Khaw, K.W., Chew, X., Al-Maatoq, M., Alharbi, R.K.: Uncovering the antecedents of trust in social commerce: an application of the non-linear artificial neural network approach. Competitiveness Rev. Int. Bus. J. **32**(3), 492–523 (2022)
39. Alnoor, A., Atiyah, A.G., Abbas, S.: Blockchain adoption strategy from the perspective of organization, technology and environment theory: a hybrid SEM and ANN approach. Glob. Bus. Rev. 09721509241255482 (2024)
40. Alnoor, A., Atiyah, A.G., Abbas, S.: Unveiling the determinants of digital strategy from the perspective of entrepreneurial orientation theory: a two-stage SEM-ANN approach. Glob. J. Flex. Syst. Manag. **25**(2), 243–260 (2024)
41. Asadi, S., Abdullah, R., Safaei, M., Nazir, S.: An integrated SEM-neural network approach for predicting determinants of adoption of wearable healthcare devices. Mob. Inf. Syst. **2019**(1), 8026042 (2019)
42. Chen, C., Tsang, S.: Predicting adoption of mobile payments from the perspective of taxi drivers. IET Intel. Trans. Syst. **13**(7), 1116–1124 (2018)
43. Loh, X.M., Lee, V.H., Hew, T.S., Lin, B.: The cognitive-affective nexus on mobile payment continuance intention during the COVID-19 pandemic. Int. J. Bank Mark. **40**(5), 939–959 (2022)
44. Ng, F.Z.X., Yap, H.Y., Tan, G.W.H., Lo, P.S., Ooi, K.B.: Fashion shopping on the go: a Dual-stage predictive-analytics SEM-ANN analysis on usage behaviour, experience response and cross-category usage. J. Retail. Consum. Serv. **65**, 102851 (2022)

45. Shahzad, F., Xiu, G., Khan, M.A.S., Shahbaz, M.: Predicting the adoption of a mobile government security response system from the user's perspective: an application of the artificial neural network approach. Technol. Soc. **62**, 101278 (2020)
46. Songkram, N., Chootongchai, S.: Adoption model for a hybrid SEM-neural network approach to education as a service. Educ. Inf. Technol. **27**(5), 5857–5887 (2022)
47. Tewari, A., Mathur, S., Srivastava, S., Gangwar, D.: Examining the role of receptivity to green communication, altruism and openness to change on young consumers' intention to purchase green apparel: a multi-analytical approach. J. Retail. Consum. Serv. **66**, 102938 (2022)
48. Tiruwa, A., Yadav, R., Suri, P.K.: Modelling Facebook usage for collaborative learning in higher education. J. Appl. Res. High. Educ. **10**(3), 357–379 (2018)
49. Wang, G., Tan, G.W.H., Yuan, Y., Ooi, K.B., Dwivedi, Y.K.: Revisiting TAM2 in behavioral targeting advertising: a deep learning-based dual-stage SEM-ANN analysis. Technol. Forecast. Soc. Chang. **175**, 121345 (2022)
50. Xiong, L., Wang, H., Wang, C.: Predicting mobile government service continuance: a two-stage structural equation modeling-artificial neural network approach. Gov. Inf. Q. **39**(1), 101654 (2022)
51. Sharma, S., Regulwar, D. G.: Prediction of evapotranspiration by artificial neural network and conventional methods. Int. J. Eng. Res, **5**, 184–187 (2016)
52. Leong, L.Y., Hew, T.S., Ooi, K.B., Tan, G.W.H.: Predicting actual spending in online group buying—an artificial neural network approach. Electron. Commer. Res. Appl. **38**(11), 100812–100898 (2019). https://doi.org/10.1016/j.elerap.2019.100898
53. Liébana-Cabanillas, F., Marinković, V., Kalinić, Z.: A SEM-neural network approach for predicting antecedents of m-commerce acceptance. Int. J. Inf. Manage. **37**(2), 14–24 (2017)
54. Ooi, K.B., Tan, G.W.H.: Mobile technology acceptance model: an investigation using mobile users to explore smartphone credit card. Expert Syst. Appl. **59**, 33–46 (2016)
55. Ooi, K.B., Lee, V.H., Tan, G.W.H., Hew, T.S., Hew, J.J.: Cloud computing in manufacturing: the next industrial revolution in Malaysia? Expert Syst. Appl. **93**, 376–394 (2018)
56. Parsad, C., Mittal, S., Krishnankutty, R.: A study on the factors affecting household solar adoption in Kerala, India. Int. J. Product. Perform. Manag. **69**(8), 1695–1720 (2020)
57. Şahin, H., Topal, B.: Impact of information technology on business performance: integrated structural equation modeling and artificial neural network approach. Sci. Iranica **25**(3), 1272–1280 (2018)
58. Sharma, S.K., Al-Badi, A., Rana, N.P., Al-Azizi, L.: Mobile applications in government services (mG-App) from user's perspectives: a predictive modelling approach. Gov. Inf. Q. **35**(4), 557–568 (2018)
59. Walle, Y.M.: Deep learning-based hybrid SEM-neural network approach for predicting mobile application adoption in whistleblowing services. Int. J. Disclo. Gov. 1–14 (2025)
60. Velleman, P.F., Wilkinson, L.: Nominal, ordinal, interval, and ratio typologies are misleading. Am. Stat. **47**(1), 65–72 (1993)
61. Alnoor, A., Salah, M., Sammar, A., Muhsen, Y. R., Sadaa, A. M., Chew, X., Bayram, G. E.: Determinants of retailers' trust in live-stream selling on social commerce platforms: Evidence using PLS-SEM and FsQCA. Int. J. Hum. Comput. Interact., 1-21 (2025)

Chapter 8
Fuzzy Set Qualitative Comparative Analysis—FsQCA

Abstract This chapter details the application of the quantitative fuzzy approach. The concept of fuzzy-set qualitative comparative analysis (fsQCA) is discussed. The chapter discusses the steps of applying fsQCA. The steps of fsQCA include data calibration (transforming data into fuzzy sets), constructing the truth tables, and boolean algebraic analysis.

Learning Objectives

1. Understand the fundamental principles of fuzzy-set qualitative comparative analysis (fsQCA).
2. Identify and apply the key steps in conducting an fsQCA.
3. Critically evaluate the strengths and limitations of using the fsQCA method in research.

8.1 Introduction

Business environment is becoming more complex and dynamic. The faster changes in the surrounding environment require business to change their strategies, procedures and practice to cope with the internal and external changes. Under the dynamic business circumstances, it is more difficult to make future prediction[s] with available statistical techniques like regression and correlation analysis. In the same way, dependency solely on the qualitative techniques cannot compensate prediction accuracy [1, 2]. This asks for more sophisticated tools and techniques for better forecasting and improved decisions. Qualitative comparative analysis (QCA) is a technique that integrates both quantitative and qualitative techniques for more asymmetric, logical and generalizable data analysis. This integration of both, makes QCA a better methodological choice as compared to the quantitative analysis that is more concerned with variance based and hypotheses testing [3]. QCA can help make outcome prediction by identifying the variety of patterns in the underlying differing circumstances. There

X. Chew et al., *Partial Least Squares Structural Equation Modeling and Complementary Methods in Business Research*, Information Systems Engineering and Management 67, https://doi.org/10.1007/978-3-032-01055-1_8

are three types of QCA including crisp set QCA, multi-value QCA, and fuzzy set QCA. Crisp set QCA is the first variant of QCA developed by the Ragin and Drass in 1980s to analyse the complex binary data to make it more logical and holistic.

Pappas and Woodside [2] described three types of QCA including crisp set QCA (csQCA), multi-value QCA (mvQCA), and fuzzy set QCA (FsQCA). CsQCA is used for the analysis of binary data complex sets. Binary data is the one with two values or can have only two states (yes or no) usually represented with 0 and 1. The purpose of csQCA is to understand the complex binary data sets where several factors lead to a conclusion usually called "multiple-conjunctural causation' [2] (p. 2) under the combination of varying complex conditions. The outcomes are non-linear and non-probabilistic with the convergence of similar outcomes under the combination of different determining factors. QCA is based on the principles of Boolean algebra and Boolean algorithms to draw 'multiple conjunctional causal relations that explain complex binary data sets in a more linear, structured, and logical way. Boolean algebra breaks down a variable into two units or components to make it more understandable and eloquent. CsQCA deals with dichotomous variables whereas mvQCA is used for multi-valued variables. The method is used to analyze the dataset in conjunction with the variables that are related to an outcome [4]. We can simply define mcQCA as a QCA technique used to analyze cases having categorical data containing more than two values. A simple example could be transcribing the traffic light (multi-value variable red, yellow, green) into three binary dummy variables i.e. light (red, yellow, green). Most simply it can be understood a QCA method which is applied when there are more than two variables but there is one outcome i.e. multi-value but one outcome as in the case of traffic lights. However, as compared to the csQCA and fsQCA, mvQCA is an under-utilized method in different disciplines [5, 6]. CsQCA has limitations in that it can only treat binary variables and hence it is not appropriate to analyse the variables that vary concerning a certain degree or level. Thus, csQCA limits using QCA methods in varying contexts and situations. FsQCA overcomes this limitation of csQCA by adopting fuzzy set and fuzzy set logic principles [7, 8]. Fuzzy set is a set whose elements have a varying degree or grade of membership and the degree or grade of variance among the elements varies between 0 and 1. Fuzzy set QCA is more reliable and widely applied because the elements of the set can take any value between 0 and 1 which is hard to capture with csQCA and mvQCA. Thus, FsQCA is superior to both csQCA and mvQCA because it allows to capture of more variations and give a deeper and more accurate understanding of the complex data under different circumstances [2, 9–11]. FsQCA is useful in identifying the different relationships that provide sufficient explanation of the desired outcome [9, 12].

Whereas the quantitative techniques are more inclined toward theory testing, and qualitative methods are more intended to theory generation (e.g. grounded theory), QCA is equally good in theory testing, theory generation, and theory elaboration [2, 13]. QCA has wider applications across different fields including organization studies, information and communication sciences, social sciences, data science, and analytics [10, 14–16]. There is an emergent interest in the 'ways of doing good research' under the notion of QCA and a good amount of scholarly work is available about the 'best practices' in doing QCA studies [8, 17]. QCA techniques are labeled a kind of mixed methodology because they integrate the tools and techniques of both quantitative and qualitative methodologies [2, 9] and benefit from the widely recognized techniques from both domains. Qualitative methods analyze the field data relevant to certain case[s] in a specific context, time, and circumstances, whereas quantitative methods make deductive reasoning by converting the data into variables and drawing the inference[s]. QCA makes a combination of both inductive and deductive (statistical and non-statistical) to make more robust predictions [9]. Drawing on the variety of benefits of qualitative analysis techniques, researchers make use of QCA to get a deeper understanding of the data mostly in quantitative studies. Whereas qualitative cases are sufficiently explained with their all dimensions and properties in a specific context, quantitative studies are more interested in explaining variables, mostly in the form of cause-and-effect relationships.

8.2 Advantages of Using FsQCA

FsQCA is used both for inductive and deductive studies providing explanation, elaboration, and test of theory[ies]. It is used either to explore the context[s] or the situation to provide a sufficient explanation of the phenomenon of interest or to test the existing assumptions and theories [12, 16]. FsQCA is also capable of performing both functions simultaneously i.e. exploration and testing of a phenomenon or developing and testing the propositions thus making it a better choice over a single method [14, 18]. FsQCA is capable of identifying the varying cases in a sample relevant to a specific proposition and providing different methodological models to explain the significantly varying outcome. It considers the contextual factors and the surrounding conditions to provide a holistic understanding of the outcome considering all the possible dimensions and properties of the case[s] under analysis. FsQCA is superior to the single variance-based analysis (for example one sample variance test because it provides alternate solutions and explanations rather than a single solution that limits the outcomes [19, 20]. FsQCA unveil the different results that can be obtained with a single data set and also can be integrated with other techniques (quantitative and qualitative) if the conditions allow [17]. As stated above FsQCA is the application of fuzzy sets that help to capture the necessary surrounding conditions to explain the outcomes and also the deficient ones to understand an outcome in its entirety [2]. These deficient or insufficient conditions are the ones about which researchers 'do not care' which implies that the outcome may or may not exist and not be involved in a specific configuration of the variables. On the other side, necessary conditions are important to reveal the underlying relationship between the factors and outcome and also the intensity of the relationship between the two. These necessary conditions may be core or peripheral ones. The first reveals a strong causal relationship while the second indicates a weaker relationship [21]. Thus, FsQCA allows practitioners to determine the nature and extent of the surrounding condition to obtain the desired results. FsQCA uses a combination of both quantitative and qualitative tools to assess and capture the level and the extent to which a variable is devoted to a set, thus it helps to overcome the deficiencies in both techniques due to which a single method cannot sufficiently capture a degree of variability in a case [7, 8]. Inherently application of FsQCA relies on binary values (0,1), however, it considers the degree of variability i.e. variable can take any value between 0 and 1. Commonly such a measure is known as a calibrated measure. Calibration usually refers to a comparison between the standard value and the measured value. Calibration helps to make accurate measurements and assures accuracy of the measurement value and instrument. However, there is less use of calibration in social sciences as compared to natural or pure sciences. In social sciences, calibration is used to interpret the variations in the outcome (qualitative) and to make comparisons between the outcomes of the cases (quantitative) [3, 22, 23]. The actual benefits of FsQCA are realized when it is preferred over a variance-based method because variance-based methods have certain limitations that FsQCA can address [11, 24, 25]. For example, one of the common limitations of variance analysis is the assumption of normal distribution. In case of abnormal distribution, the variance analysis cannot be applied. Whereas FsQCA is capable of capturing both normal and abnormal distribution. Similarly, variance analysis determines the net effect of the competing variables over the outcome variable, but FsQCA measures the complex and asymmetrical relationship between the outcome and predictors. For example, in studies that consider respondents' demographics (age, education, gender, etc.) as control variables and exclude their effect from the outcome while using generic variance-base methods, FsQCA accounts for these variations in the explanation of the outcome[s]. FsQCA works on the principle that variation in the outcome may be the result of the varying and different combinations of the situation or the attributes and every combination accounts differently towards the outcome.

Another advantage of FsQCA is that it can be applied to different sample sizes ranging from smaller to larger ones. The sample size could be as small as less than 50 to as large as 1000s. The size of the sample allows researchers to explain the results separately or to detect the patterns across the cases [26]. In the same way, FsQCA is capable of considering and treating different data types, for example, the data obtained through the Likert scale, through clicks/options, or multidimensional data. For this, the data is first transformed into fuzzy sets. Also, FsQCA can deal with categorical data which do not need to be converted into fuzzy sets. To use categorical data, it needs to be in the form of binary data (obtained via dichotomous scale) or could take the range between 0 and 1 (data obtained through numerical or semantic differential scale). Another benefit of using FsCA is that it also captures all possible variables that influence the outcome in a causal relationship. The FsQCA technique allows to combination of the variables that are not taken into account while using a variance-based method because the variance-based method only accounts for the variable that causes the main effect on the outcome and not all the variables [2, 11]. Proponents of the variance-based method argue that other variables (e.g. control variables) are not significant in the majority of the cases and hence are omitted from the final results. Whereas the variance-based methods are only focused on the sample as a whole single unit, FsQCA divided the sample into smaller multiple units with different characteristics to facilitate the analysis under different combinations and conditions. In simple words, we can say that while the variance-based methods do not consider 'outliers' the FsQCA takes into account the 'outliers' for comprehensive and robust results. The configuration of the sample constitutes a 'sub-set' of the sample and each configuration denotes one sub-set. This is not uncommon that the outlier[es] is[are] present in the case of only a few sub-sets and not in all. Hence, we can say that in some cases FsQCA aligns with the variance-based method when the outliers are absent, while in some cases FsQCA explains the cases where there are outliers [2]. In this way, it is possible that all the representatives in a sample do not necessarily influence all outcomes making it a superior methodological choice over the variance-based methods [21, 25]. In another way we can say that FsQCA does not need to always consider the outliers, however, it depends on the sample configuration. FsQCA is more useful for examining contrarian cases. In statistical terms, contrarian cases are those which deviate from the usual pattern or where there exists unexpected relationship[s] between the variables. In such cases, FsQCA helps identify the outliers and indicates that there exist several cases in the sample (sub-set) that cannot be explained through the main effect {e.g. relationship between the predictors and the outcome[s]). It is commonly found that many of the research articles do not consider the existence of contrarian cases in the tested relationship and outcome.

Another added advantage of using FsQCA is that it explores both positive and negative relationship between the variables under the combination of the different conditions. This make this FsQCA a preferred choice over the simple regression analysis which only accounts for the existence of [in]significant relationship between the variables usually under controlled conditions, For example, experimental studies [27]. In the same way, regression analysis is concerned with the existence or non-existence of symmetrical relationship between the variables (symmetrical vs asymmetrical) under certain conditions. However, it must be realized that the similar relationship may exist under the different conditions, which regression analysis cannot capture. FsQCA considers the co-existence of the different conditions influencing the causal relationship [19].

8.2.1 *Fuzzy Set Qualitative Comparative Analysis: Business Research Applications*

In many instances, businesses need to make objective decisions which often needs quantification of the variables, Typically, in business studies, focus is more on the adoption of the quantitative methods using statical techniques and specifying the surrounding factors. Thus, the results or the outcomes obtained have limited applications. As stated above that consideration of the static conditions may hamper the prediction accuracy in the wake of changing business circumstances [2, 28]. This must be realized that market conditions are always changing and thus the environment in which a business operates is continuously becoming complex, ambiguous and inconsistent. Business can no longer rely on the historical data to make the predictions due to the changing future circumstances [29]. Earlier studies in the fields of business and management have reported similar instances where the dynamic business environment is an important factor to make the future predictions (e.g. Refs. [30–32]). Several scholars have argued that the application of pure statistical techniques, for example structural equation modelling (SEM) May not be an appropriate choice and misleading in reporting the results under the specific conditions [33, 34]. This caused researchers to consider other to opt for the other methods such as FsQCA which can better explain the complex business scenario [35]. Kumar et al. [1] (p. 179) argue that earlier investigations involving the complex and dynamic situations have paved the way for 'complex adaptive systems' leading to the applications of "complexity theory to scientific investigation" with the adoption of novel methods such as FsQCA. FsQCA has been recognized a much appreciable tool to make closer to real prediction, it is more objective, more reliable [36–38]. In comparison to the other statistical method which generate positive or negative outcomes, FsQCA provide an additional advantage of predicting no outcome under the combination of varying business conditions [39]. The traditional statistical methods (e.g. test of hypotheses for significance) are becoming obsolete and outdated. More attention is being paid to the discovery of novel techniques for better and improved prediction accuracy. FsQCA is one of such techniques that has emerged as a one sophisticated tool to investigate the complex business framework[s] and to better anticipate the desired outcome[s] [2, 40]. A noticeable work of Pappas and Woodside [2] have led to the growing interest of the business researcher in employing FsQCA technique. FsQCA compensates the limitations of both quantitative and qualitative and quantitative techniques by application of qualitative methods for capturing the contextual factors for deeper understanding of the phenomenon of an interest in combination with the quantitative methods to investigate the phenomenon across large sample size thus offering 'generalization of the results [41, 42].

FsQCA offers more depths and breadth because it encompasses all the three 'favorable, non-favoravle and non-existing factors in the comprehension of the results. It also specifies a condition[s] if the desired outcome cannot be obtained in the absence of such condition[s]. If no additional condition[s] is[are] required that the existing one is considered as a sufficient condition as it generates the desired results [40, 43, 44]. The prime concern of FsQCA is determining the combination of the conditions or factors essentially needed to achieve the required outcome[s] and to indicate that which these combinations are similar with respect to a specific set of requirements [1, 45, 46]. Generally, FsQCA generates three set of findings, first indication of variables that are related to each other in one causal relationship but are not related or negatively associated in other relationship thus reflecting an asymmetrical relationship between the variables [36, 47]. Second is the equifinality of the findings i.e. similar outcomes may be generated with the different existing conditions [43, 48]. In third instance, FsQCA demonstrates that the combined effect of the predictors on the outcome. It depends on the predictors are integrated instead of determining the net effect of a single predictor i.e. conjunctural causation. Conjunctural causation means that events are the results of a chain of multiple intersecting forces/factors/conditions [49–51]. Prentice [40] and Woodside et al. [52] argued that because the relationship between the outcome and the predictors reported to be asymmetrical by the previous studies using symmetrical techniques (SEM, PLS-SEM) under the established conditions, there is need to re-evaluate those results using asymmetrical techniques such as FsQCA. The other point of using asymmetrical techniques (FsQCA) is because real-world dataset(situations) does not exhibit symmetry that can lead to the biasness issue in result estimation if not addressed properly. Kumar et al. [1] reported that they searched Scopus database for articles relevant to the business and management studies using prominent key words under the umbrella of symmetrical techniques (SEM, PLS-SEM etc.) then they turned up with more than two million articles. This necessitated the application of asymmetrical techniques to obtain the more relevant articles. This is quite understandable as many scholars argue that relying on older hypotheses techniques (e.g. null hypothesis significance testing) cannot generate the robust results anymore due to the emergence of complex and dynamic business environment [1, 40, 52]. Kumar et al. [1] (p. 179) recommended that FsQCA is more reliable technique to predict the outcome for the business in emerging environment which is "disruptive, volatile, uncertain, complex, and ambiguous (DVUCA), hence any previous study[ies] that used symmetrical techniques must be revisited and reevaluated to affirm the outcomes [31, 32, 53]. This is especially true for the previous studies that used the lens of complexity theory. Complexity theory proposes that systems are unpredictable, confined within order generating rules. It believes in the existence of heterogeneity within organizational sub-systems and the way sub-systems influence emergent behaviour and the outcomes.

8.3 How to Use FsQCA—A Step-By-Step Process

For implementation of the FsQCA method, one must have a deeper understanding of the variables to be studied and the underlying theory[ies]. Additionally, the knowledge and understanding of the context will add further to generate and robust results. This is important for conversion of the data into fuzzy sets, simplification of the various solution offered by FsQCA and interpretation of the results. While using the FsQCA, researchers must consider the limitations of using the methods to avoid the personal biases (both from the researcher and the respondents). A comprehensive understanding of the methods and its application can help to develop a deeper and all-inclusive understanding of the results. Traditional mixed methods can be used either way (i.e. sequential or parallel) depending on the nature of the study. For example, when the context is novel, it is appropriate to use qualitative methods first to develop understanding of the context and the research problems for the development of propositions. After that the quantitative methods are employed to converge the results. FsQCA is meant identifying the complex and logical relationship between the predictors (conditions) to derive the overarching results instead of focusing on the single causal effect of the individual variable. While some of the scholars may argue over the calculation of qualitative data but FsQCA has made easy calibration of qualitative data (interviews, documents) into fuzzy sets. There are numerous studies which have applied FsQCA for the same purpose (e.g. Refs. [10, 54, 55]). FsQCA is being used across the field to analyse both quantitative and qualitative data under the umbrella of one study [56]. Integration of the quantitative and qualitative methods provide researchers with a comprehensive methodology to infer more reliable and trustworthy results. Pappas and Woodside [2] argue that one must be aware of the common method bias while using the FsQCA. This happens because the data is derived from the same source both for the predictors and the outcome. Commonly, Herman's one-factor test is employed to detect the common method bias. Similarly, another limitation of the FsQCA method is that it is relatively weaker in providing the solutions/results that can be generalized. Though it provides complex and detailed results but those are not generalizable [27]. Another problem which researchers may face is to ensure the validity that can permit the replications. For this, researchers must be attuned to the guidelines and conscious decisions while applying the method. However, it must be noted that construct validity and reliability are not direct subject of the FsQCA rather they relate more to the nature of the variables being used in the study. They relate to the 'construct[s] not to the method [2].

As noted above, FsQAC captures all the necessary conditions, integrating quantitative and qualitative methods to generate the outcomes. It is noted that FsQCA explains the variance by transforming the data into a range between zero and one. Earlier studies of Elshaer et al. [57] and Xu et al. [58] necessitated the need to develop the configuration of the conditions for a deeper understanding of the underlying relationships. In one of our earlier studies, to unveil the optimal configuration of impulsive buying behavior, we used FsQCA to explain the different surrounding conditions that explain the buyer's impulsive buying behavior. In doing so, multiple sets of varying conditions led to the study's outcome [59]. We performed a contrarian case analysis to determine the unexplained effects (not determined by the predictors of the study). This is one of the necessary conditions to use the FsQCA method. We segregated the sample into five groups of equal size for cross-tabulation and calculated the relationship between the variables using correlation measures. It was found that each set of samples showed different correlation measures between the variables and hence it required configuration analysis. The next step in the FsQCA is to calibrate the data. For this purpose, the data can be converted into the different values between 0 and 1. We considered 0.95, 0.5, and 0.05 as breakpoints to a log-odds metric range (0–1). Following the data calibration (converting the data into values between 0 and 1), the next step is to perform the truth table analysis. This is done with the help of algorithmic techniques with the identification of predictors and outcome variables. Predictors determine the causal conditions that determine the effect. Configuration analysis was also performed to cover the other conditions (underlying conditions) that also contributed to the outcome (impulsive behavior in the case of our study). This is a mandatory step to do in case of coverage value > 0.5 and consistency value > 0.9 as proposed by XinYing et al. [60]. FsQCA provides a number of solutions and thus helps to choose the best solution, in the case of our study we generated 15 alternate solutions combining the different underlying conditions (i.e. Variables) (see Alnoor et al. [59]).

As mentioned above FsQCA is an extension of QCA originally developed by Ragin [7]. The technique was initially suited to the studies, especially in the broader fields of political, social, and organizational studies with small sample sizes. Later on, scholars started to use in other sub-fields like business and management, entrepreneurship with medium and large samples [11, 61–63]. Recently, FsQCA is being used widely across social and business domains due to its inherent feature of 'bridging the gap' between the two i.e. qualitative and quantitative [3, 22]. The techniques suites more to the conditions when there are multiple causes i.e. the outcome is caused by different conditions i.e. identification of the various paths that lead to the same outcome [64]. The main intention behind use of FsQCA is identification of the sufficient conditions to explain the outcome [19]. According to Stokke [65] (p. 510) argued that FsQCA strengthen the methodological middle ground between qualitative and quantitative analysis. In business and management studies (especially the case study approach), FsQCA is preferred choice as it allows for more generalization across the contexts with improved accuracy of the results [66].

FsQCA is a balanced approach between the quantitative and qualitative methods that transforms the variables into a continuous degree of "belonging" and "membership" [67] (p. 226). FsQCA explains the complexity at case levels and generalizability across the cases, especially when the causal relationship is complex and cannot be explained through 'net effect [68, 69]. It helps identifying the most relevant factor[s] that lead to the desirable outcome and the way to develop a combination of the conditions for a comprehensive understanding of the other complementary and substitute factors [21, 70]. In the following we briefly explain the steps involved using the FsQCA in general and in business and management studies in particular. FsQCA involves three steps i.e. data calibration (transforming the data into fuzzy sets), construction of truth tables, and fuzzy set analysis using Boolean algebra approach.

8.3.1 Data Calibration (Transforming Data into Fuzzy Sets)

In the first step, the data is converted to the fuzzy sets where the values of variables are operationalized in terms of membership scores [3, 22, 71]. This is called data calibration. Fuzzy set membership scores are dependent on the extent to which a cases is related to a fuzzy set. Every combination of distinctive conditions is described as having certain characteristics. The data files are imported from the software (stata, spss) that must be 'fuzzified' [67]. In the majority of the cases, the data is generated/collected through a Likert scale. Different sets of cases are developed, and in each set of cases variables are transformed on the basis of degree of membership (ranging 0–1) to create the score where 0 denotes no membership and 1 denotes full membership, whereas 0.5 refers to the crossover point i.e. maximum ambiguity in the conditions. Researchers must possess a comprehensive essential theoretical knowledge for specifying the three "qualitative breakpoints" i.e. full membership, no membership, and the crossover point [72] (p. 18). Researchers must rationalize while refereeing membership to a specific breakpoint. This is important for aligning the values across fuzzy set and Likert scale [70]. Based on the membership score, sub-sets are analyzed for obtaining consistency and coverage across the fuzz sets [3, 22]. Consistency refers to theoretical consistency of the set i.e. the extent to what extent the conditions and outcomes across the sub-set are identical. This also indicates the cases sharing a single condition or a combination of conditions. Similarly, theoretical coverage denotes the relevancy of the condition[s] to the outcome[s] i.e. the extent to which the observed cases have been covered under the defining relationship where a lower degree of coverage means that several combination of paths to the same outcome [72–74]. For the easy conversion of data into fuzzy set it is recommended to use the FsQCA program software that is designed out of the procedure suggested by Ragin [3, 22].

The second step after the calibration is construction of truth tables. For this purpose, fuzzy set scores are used (from step 1) to develop the data matrix as a truth table with 2k rows (2k is configuration of conditions, k is no. of causal conditions for an outcome) for Boolean algebra analysis. This is basically used to check if the proposed combination is true or false for an outcome. The truth table measures the impact of each cause in the underlying conditions/relationships. It needs a detailed analysis to determine the effect of all causes or causal conditions with different combinations of the conditions [75]. However, in case of smaller samples, large configuration is not feasible where some of the rows contain zero or remainder case[s]. In such cases, truth table can be further improved through the combination of conditions. The relevancy of the conditions is determined through the number of cases that exceed the crossover point value of 0.5 in each case via establishing the threshold level [3, 22].

8.3.2 Boolean Algebraic Analysis

FsQCA can analyze the different combinations of conditions at the same time. To minimize the number of combinations algorithms are used. With the specification of the frequency and consistency threshold the number of truth tables is reduced [70, 73]. Frequency threshold is used to determine the relevance of the combination of the conditions and its value range from 1 to 2 depending on the relative sample size [3, 22]. Similarly, the frequency threshold suggest the combination which score high in an outcome, a score above 0.75 is considered high level of consistency. Consistency threshold helps to isolate the irrelevant conditions and thus proposed critical conditions for an outcome [71]. The analysis stage identifies the most relevant conditions for an outcome and limits the number of solutions [76].

References

1. Kumar, S., Sahoo, S., Lim, W.M., Kraus, S., Bamel, U.: Fuzzy-set qualitative comparative analysis (fsQCA) in business and management research: a contemporary overview. Technol. Forecast. Soc. Chang. **178**, 121599 (2022)
2. Pappas, I.O., Woodside, A.G.: Fuzzy-set qualitative comparative analysis (fsQCA): guidelines for research practice in Information Systems and marketing. Int. J. Inf. Manage. **58**, 102310 (2021)
3. Ragin, C.C.: Redesigning Social Inquiry: Fuzzy Sets and Beyond, vol. 240. Wiley Online Library (2008)
4. Cronqvist, L., Berg-Schlosser, D.: Multi-value QCA (mvQCA). In: Configurational Comparative Methods: Qualitative Comparative Analysis (QCA) and Related Techniques, pp. 69–86 (2009)
5. Thiem, A., Dusa, A.: QCA: a package for qualitative comparative analysis. R J. **5**(1), 87 (2013)
6. Vink, M.P., Van Vliet, O.: Not quite crisp, not yet fuzzy? Assessing the potentials and pitfalls of multi-value QCA. Field Methods **21**(3), 265–289 (2009)
7. Ragin, C.C.: Fuzzy-set social science. University of Chicago Press (2000)
8. Rihoux, B., Ragin, C.C. (eds.): Configurational Comparative Methods: Qualitative Comparative Analysis (QCA) and Related Techniques, vol. 51. Sage
9. Ordanini, A., Parasuraman, A., Rubera, G.: When the recipe is more important than the ingredients: a qualitative comparative analysis (QCA) of service innovation configurations. J. Serv. Res. **17**(2), 134–149 (2014)
10. Pappas, I.O., Kourouthanassis, P.E., Giannakos, M.N., Chrissikopoulos, V.: Explaining online shopping behavior with fsQCA: the role of cognitive and affective perceptions. J. Bus. Res. **69**(2), 794–803 (2016)
11. Woodside, A.G.: Embrace perform model: complexity theory, contrarian case analysis, and multiple realities. J. Bus. Res. **67**(12), 2495–2503 (2014)
12. Greckhamer, T., Furnari, S., Fiss, P.C., Aguilera, R.V.: Studying configurations with qualitative comparative analysis: best practices in strategy and organization research. Strateg. Organ.. Organ. **16**(4), 482–495 (2018)
13. Misangyi, V.F., Greckhamer, T., Furnari, S., Fiss, P.C., Crilly, D., Aguilera, R.: Embracing causal complexity: the emergence of a neo-configurational perspective. J. Manag.Manag. **43**(1), 255–282 (2017)
14. Pappas, I.O., Papavlasopoulou, S., Mikalef, P., Giannakos, M.N.: Identifying the combinations of motivations and emotions for creating satisfied users in SNSs: an fsQCA approach. Int. J. Inf. Manage. **53**, 102128 (2020)
15. Park, Y., Mithas, S.: Organized complexity of digital business strategy: a configurational perspective. MISQ **44**(1) (2020)
16. Park, Y., Fiss, P.C., El Sawy, O.A.: Theorizing the multiplicity of digital phenomena: the ecology of configurations, causal recipes, and guidelines for applying QCA. Manage. Inf. Syst. Quart. **44**, 1493–1520 (2020)
17. Schneider, C.Q., Wagemann, C.: Standards of good practice in qualitative comparative analysis (QCA) and fuzzy-sets. Comp. Sociol.Sociol. **9**(3), 397–418 (2010)
18. Pappas, I.O.: User experience in personalized online shopping: a fuzzy-set analysis. Eur. J. Mark. **52**(7/8), 1679–1703 (2018)
19. Woodside, A.G.: Moving beyond multiple regression analysis to algorithms: calling for adoption of a paradigm shift from symmetric to asymmetric thinking in data analysis and crafting theory. J. Bus. Res. **66**(4), 463–472 (2013)
20. Woodside, A.G.: The good practices manifesto: overcoming bad practices pervasive in current research in business. J. Bus. Res. **69**(2), 365–381 (2016)
21. Fiss, P.C.: Building better causal theories: a fuzzy set approach to typologies in organization research. Acad. Manag. J.Manag. J. **54**(2), 393–420 (2011)
22. Ragin, C.C.: Redesigning Social Inquiry: Fuzzy Sets and Beyond. The University of Chicago Press, Chicago (2008)

23. Vis, B.: The comparative advantages of fsQCA and regression analysis for moderately large-N analyses. Sociol. Methods Res. **41**(1), 168–198 (2012)
24. El Sawy, O.A., Malhotra, A., Park, Y., Pavlou, P.A.: Research commentary—seeking the configurations of digital ecodynamics: it takes three to tango. Inf. Syst. Res. **21**(4), 835–848 (2010)
25. Liu, Y., Mezei, J., Kostakos, V., Li, H.: Applying configurational analysis to IS behavioural research: a methodological alternative for modelling combinatorial complexities. Inf. Syst. J. **27**(1), 59–89 (2017)
26. Greckhamer, T., Misangyi, V.F., Fiss, P.C.: Chapter 3 the two QCAs: from a small-N to a large-N set theoretic approach. In: Configurational Theory and Methods in Organizational Research, pp. 49–75. Emerald Group Publishing Limited
27. Gligor, D., Bozkurt, S.: FsQCA versus regression: the context of customer engagement. J. Retail. Consum. Serv.Consum. Serv. **52**, 101929 (2020)
28. Tiberius, V., Rasche, C.: Prognosemarkte. ¨ Zeitschrift für Plan. Unternehmenssteuerung **21**, 467–472 (2011)
29. Woodside, A.G.: Have your cake and eat it too: achieving scientific legitimacy. Ind. Mark. Manage. **69**, 53–61 (2018)
30. Kumar, A., Shankar, R., Aljohani, N.R.: A big data driven framework for demand-driven forecasting with effects of marketing-mix variables. Ind. Mark. Manage. **90**, 493–507 (2020)
31. Prommer, L., Tiberius, V., Kraus, S.: Exploring the future of startup leadership development. J. Bus. Ventur. InsightsVentur. Insights **14**, e00200 (2020)
32. Tiberius, V., Hauptmeijer, R.: Equity crowdfunding: forecasting market development, platform evolution, and regulation. J. Small Bus. Manage. **59**(2), 337–369 (2021)
33. Khedhaouria, A., Cucchi, A.: Technostress creators, personality traits, and job burnout: a fuzzy-set configurational analysis. J. Bus. Res. **101**, 349–361 (2019)
34. Papatheodorou, A., Pappas, N.: Economic recession, job vulnerability, and tourism decision making: a qualitative comparative analysis. J. Travel Res. **56**(5), 663–677 (2017)
35. De Toni, A.F., Pessot, E.: Investigating organisational learning to master project complexity: an embedded case study. J. Bus. Res. **129**, 541–554 (2021)
36. Lee, C.K.H.: How guest-host interactions affect consumer experiences in the sharing economy: new evidence from a configurational analysis based on consumer reviews. Decis. Support. Syst.. Support Syst. **152**, 113634 (2022)
37. Olya, H., Taheri, B., Farmaki, A., Joseph Gannon, M.: Modelling perceived service quality and turnover intentions in gender-segregated environments. Int. J. Consum. Stud.Consum. Stud. **46**(1), 200–217 (2022)
38. Thornton, S.C., Henneberg, S.C., Leischnig, A., Naudé, P.: It's in the mix: how firms configure resource mobilization for new product success. J. Prod. Innov. Manag.Innov. Manag. **36**(4), 513–531 (2019)
39. Huarng, K.H., Yu, T.H.K.: Using qualitative approach to forecasting regime switches. Qual. Quant. **51**, 2035–2048 (2017)
40. Prentice, C.: Testing complexity theory in service research. J. Serv. Mark. **34**(2), 149–162 (2020)
41. Acquah, I.S.K., Naude, M.J., Sendra-García, J.: Supply chain collaboration in the petroleum sector of an emerging economy: comparing results from symmetrical and asymmetrical approaches. Technol. Forecast. Soc. Chang. **166**, 120568 (2021)
42. Dahms, S.: Foreign-owned subsidiary knowledge sourcing: the role of location and expatriates. J. Bus. Res. **105**, 178–188 (2019)
43. Kopplin, C.S., Rösch, S.F.: Equifinal causes of sustainable clothing purchase behavior: an fsQCA analysis among generation Y. J. Retail. Consum. Serv.Consum. Serv. **63**, 102692 (2021)
44. Küçükergin, K.G., Çalışkan, C., Dedeoglu, B.B., Birinci, M.C.: Analyzing the role of constraints and motivations behind traveling in the prediction of destination choice: evidence from PLS-SEM and fsQCA. Int. J. Tour. Res. **23**(6), 1191–1209 (2021)
45. Bandara, R., Fernando, M., Akter, S.: Managing consumer privacy concerns and defensive behaviours in the digital marketplace. Eur. J. Mark. **55**(1), 219–246 (2021)

46. Timmer, S., Kaufmann, L.: Conflict minerals traceability—a fuzzy set analysis. Int. J. Phys. Distrib. Logist. Manag.Distrib. Logist. Manag. **47**(5), 344–367 (2017)
47. Gligor, D.M., Golgeci, I., Rego, C., Russo, I., Bozkurt, S., Pohlen, T., et al.: Examining the use of fsQCA in B2B marketing research: benefits, current state and agenda for future research. J. Bus. Ind. Mark. **37**(7), 1542–1552 (2022)
48. de Guinea, A.O., Raymond, L.: Enabling innovation in the face of uncertainty through IT ambidexterity: a fuzzy set qualitative comparative analysis of industrial service SMEs. Int. J. Inf. Manage. **50**, 244–260 (2020)
49. Cheng, C.F., Chang, M.L., Li, C.S.: Configural paths to successful product innovation. J. Bus. Res. **66**(12), 2561–2573 (2013)
50. Mikalef, P., Boura, M., Lekakos, G., Krogstie, J.: Big data analytics and firm performance: Findings from a mixed-method approach. J. Bus. Res. **98**, 261–276 (2019)
51. Romero-Castro, N., Piñeiro-Chousa, J., Pérez-Pico, A.: Dealing with heterogeneity and complexity in the analysis of the willingness to invest in community renewable energy in rural areas. Technol. Forecast. Soc. Chang. **173**, 121165 (2021)
52. Woodside, A.G., Megehee, C.M., Isaksson, L., Ferguson, G.: Consequences of national cultures and motivations on entrepreneurship, innovation, ethical behavior, and quality-of-life. J. Bus. Ind. Marketing **35**(1), 40–60 (2020)
53. Gassmann, S.E., Nunkoo, R., Tiberius, V., Kraus, S.: My home is your castle: forecasting the future of accommodation sharing. Int. J. Contemp. Hosp. Manag.Manag. **33**(2), 467–489 (2021)
54. Basurto, X., Speer, J.: Structuring the calibration of qualitative data as sets for qualitative comparative analysis (QCA). Field Methods **24**(2), 155–174 (2012)
55. Henik, E.: Understanding whistle-blowing: a set-theoretic approach. J. Bus. Res. **68**(2), 442–450 (2015)
56. Kane, H., Lewis, M.A., Williams, P.A., Kahwati, L.C.: Using qualitative comparative analysis to understand and quantify translation and implementation. Trans. Beha. Med. **4**(2), 201–208 (2014)
57. Elshaer, I.A., Alrawad, M., Lutfi, A., Azazz, A.M.: Social commerce and buying intention post COVID-19: evidence from a hybrid approach based on SEM–fsQCA. J. Retail. Consum. Serv.Consum. Serv. **76**, 103548 (2024)
58. Xu, X., Jia, Q., Tayyab, S.M.U.: Exploring the dual routes in influencing sales and adoption in augmented reality retailing: a mixed approach of SEM and FsQCA. Internet Res. (2024)
59. Alnoor, A., Abbas, S., Khaw, K.W., Muhsen, Y.R., Chew, X.: Unveiling the optimal configuration of impulsive buying behavior using fuzzy set qualitative comparative analysis and multi-criteria decision approach. J. Retail. Consum. Serv.Consum. Serv. **81**, 104057 (2024)
60. XinYing, C., Tiberius, V., Alnoor, A., Camilleri, M., Khaw, K.W.: The dark side of metaverse: a multi-perspective of deviant behaviors from PLS-SEM and fsQCA findings. Int. J. Hum.–Comput. Interaction, 1–21 (2024)
61. Cooper, B., Glaesser, J.: Using case-based approaches to analyse large datasets: a comparison of Ragin's fsQCA and fuzzy cluster analysis. Int. J. Soc. Res. Methodol.Methodol. **14**(1), 31–48 (2011)
62. García-Castro, R., Aguilera, R.V., Ariño, M.A.: Bundles of firm corporate governance practices: a fuzzy set analysis. Corp. Governance: Int. Rev. **21**(4), 390–407 (2013)
63. Rihoux, B., Álamos-Concha, P., Bol, D., Marx, A., Rezsöhazy, I.: From niche to mainstream method? A comprehensive mapping of QCA applications in journal articles from 1984 to 2011. Political Res. Quart. 175–184 (2013)
64. Mendel, J.M., Korjani, M.M.: Theoretical aspects of fuzzy set qualitative comparative analysis (fsQCA). Inf. Sci. **237**, 137–161 (2013)
65. Stokke, O.S.: Qualitative comparative analysis, shaming, and international regime effectiveness. J. Bus. Res. **60**(5), 501–511 (2007)
66. Woodside, A.G.: Bridging the chasm between survey and case study research: research methods for achieving generalization, accuracy, and complexity. Ind. Mark. Manage. **39**(1), 64–75 (2010)

67. Kent, R.: Cases as configurations: using combinatorial and fuzzy logic to analyse marketing data. Int. J. Mark. Res. **47**(2), 205–228 (2005)
68. Aguilera-Caracuel, J., Fedriani, E.M., Delgado-Márquez, B.L.: Institutional distance among country influences and environmental performance standardization in multinational enterprises. J. Bus. Res. **67**(11), 2385–2392 (2014)
69. Crilly, D.: Predicting stakeholder orientation in the multinational enterprise: a mid-range theory. J. Int. Bus. Stud. **42**, 694–717 (2011)
70. Chang, M.L., Cheng, C.F.: How balance theory explains high-tech professionals' solutions of enhancing job satisfaction. J. Bus. Res. **67**(9), 2008–2018 (2014)
71. Meuer, J.: Archetypes of inter-firm relations in the implementation of management innovation: a set-theoretic study in China's biopharmaceutical industry. Organ. Stud. **35**(1), 121–145 (2014)
72. Kraus, S., Ribeiro-Soriano, D., Schüssler, M.: Fuzzy-set qualitative comparative analysis (fsQCA) in entrepreneurship and innovation research—the rise of a method. Int. Entrep. Manag. J.Entrep. Manag. J. **14**, 15–33 (2018)
73. Ragin, C.C.: Qualitative comparative analysis using fuzzy sets (fsQCA). In: Configurational Comparative Methods: Qualitative Comparative Analysis (QCA) and Related Techniques, pp. 87–122. (2009)
74. Schneider, M.R., Eggert, A.: Embracing complex causality with the QCA method: an invitation. JBM-J. Bus. Market Manage. **7**(1), 312–328 (2014)
75. Woodside, A.G., Baxter, R.: Achieving accuracy, generalization-to-contexts, and complexity in theories of business-to-business decision processes. Ind. Mark. Manage. **42**(3), 382–393 (2013)
76. Mas-Verdú, F., Ribeiro-Soriano, D., Roig-Tierno, N.: Firm survival: the role of incubators and business characteristics. J. Bus. Res. **68**(4), 793–796 (2015)

Chapter 9
Outline of ANFIS Analysis

Abstract This chapter reviews one of the most important statistical methods for complex and nonlinear relationships. Adaptive Neuro-Fuzzy Inference Systems (ANFIS) is a modern and reliable methodology with a five-layer structure. This methodology addresses issues in diverse fields, such as medicine, engineering, education, and the social sciences. This chapter also discusses the integration of ANFIS and PLS-SEM in addressing study hypotheses and predicting the most important causes of phenomena. The chapter relies on review research included in prestigious databases to demonstrate the practical aspect. Therefore, the chapter discusses the importance of the ANFIS method, resulting from its ability to improve the accuracy of results, monitor complex interactions, and enhance decision-making processes.

Learning Objectives

1. A review of contemporary studies aims to define adaptive neuro-fuzzy inference systems (ANFIS) and their essential role in data modeling and prediction.
2. Clarify the process of mixing neural networks and fuzzy logic in ANFIS to understand how to address complex and nonlinear problems.
3. Predicting the accuracy of ANFIS using RMSE and considering its limitations and new treatments in the light of classical statistical methods.

9.1 Introduction

Artificial intelligence (AI) is the trend that has transformed the business world and the global landscape of robotics and digital transformation. AI technologies have provided numerous tools for solving complex problems. AI consists of several branches, including neural networks, fuzzy modeling, simulation, genetic algorithms, and others. AI technologies provide more accurate predictions, contributing to the provision of accurate information to stakeholders and policymakers for reliable decision-making [1]. Many practitioners adopt computing and manufacturing

X. Chew et al., *Partial Least Squares Structural Equation Modeling and Complementary Methods in Business Research*, Information Systems Engineering and Management 67, https://doi.org/10.1007/978-3-032-01055-1_9

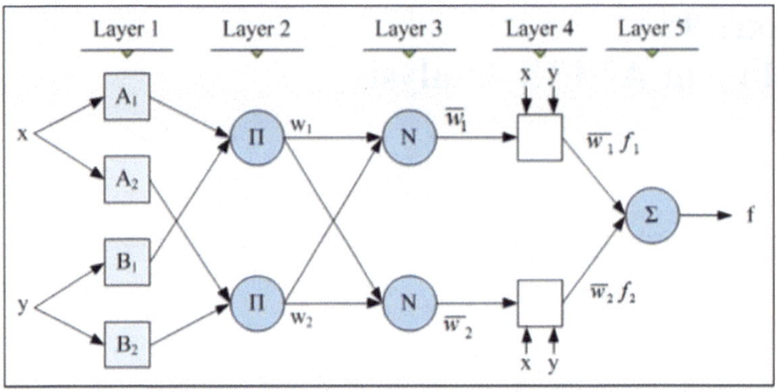

Fig. 9.1 ANFIS process

moderation. Funny models have gained great importance in modeling operations, from traditional operations such as slope to modern operations such as AI technologies [2]. AI technologies have contributed to improving the statistical tools applied in various sciences, including social sciences. The systems and applications were used for AI in different sciences, including medicine, economy, engineering, marine, military fields, and others. The Adaptive Neuro-Fuzzy Investment System (ANFIS) is one of the emerging applications that have been used with structure equation modeling (SEM) [3, 4]. ANFIS is AI technologies that combine interpretation of mysterious utility systems and learning capabilities of nerve networks. The ANFIS curriculum merges the advantages of nervous network technologies and mysterious logic. By combining nervous networks, abilities and mysterious fuzzy logic methods, complex problems can solve and make confidence -worth predictions [5]. Figure 9.1 reveals the work of the ANFIS that combines the artificial neural network and fuzzy investment systems.

Figure 9.1 reveals that an ANFIS model consists of three models as in the artificial neural networks model, which is an input, hidden and output layer. In addition to that there is x and y represent fuzzy inference systems. ANFIS combines neural network learning capabilities with system interpretation and fuzzy inference technologies (ANFIS). This technique tries to reduce uncertainty and capture non-linear interactions by combining neural network learning techniques and the benefits of fuzzy logic. The literature recommends the ANFIS technique for solving complicated issues [3, 4]. Equations 9.1 and 9.2 show the mechanism for applying the first-order fuzzy model.

$$\text{If "x" is "}A_1\text{" and "y" is "}B_1\text{" then "}z_1 = p_1x + q_1y + r_1\text{"} \tag{9.1}$$

$$\text{If "x" is "}A_2\text{" and "y" is "}B_2\text{" then "}z_2 = p_2x + q_2y + r_2\text{"} \tag{9.2}$$

where "x" and "y" are non-fuzzy inputs, A_i and B_i are fuzzy sets, z_i is the output value; p_i, q_i and r_i are parameters of output function. There are five layers of ANFIS analysis.

Layer 1: O_1; i represents the output of node i in layer l (Eqs. 9.3 and 9.4), and O_1; i is the membership function of A_i and B_{i-3}. Every node i in layer 1 is an adaptive unit with a function given by:

$$O_{1,i} = \mu_{A_i}(x), \text{ for } i = 1, 2, 3 \tag{9.3}$$

$$O_{1,i} = \mu_{Bh_3}(y), \text{ for } i = 4, 5, 6 \tag{9.4}$$

where x and y are inputs, and A_i and B_{i-3} are membership functions for inputs.

Layer 2: The nodes in this layer are fixed nodes labelled Π, which multiply the incoming inputs (Eq. 9.5):

$$O_{2,i} = w_i = \mu_{A_i}(x).\mu_{B_i}(y), \ i = 1, 2 \tag{9.5}$$

Layer 3: The fixed nodes of this layer are called N. In this layer, the firing force of the rules is calculated to the sum of the force of the rules (Eq. 9.6):

$$O_{3,i} = \overline{w}_i = w_i/(w_1 + w_2), \ \text{i} = 1, 2 \tag{9.6}$$

Layer 4: This layer includes a set of linear functions (Eq. 9.7):

$$O_{4,i} = \overline{w}_{i}. \cdot f_i = \overline{w}_i \cdot (p_i X + q_i y + r_i) \tag{9.7}$$

where w_i is an output from layer 3, and p_i, q_i, r_{ig} is the consequent parameter set.

Layer 5: In this layer, the outputs of the fourth layer are collected in the layer called P (Eq. 9.8):

$$O_{5,i} = \text{overall output} = \sum_i \overline{w}_i f = \frac{\sum_i w_i f_i}{\sum_i w_i} \tag{9.8}$$

RMSE evaluates the accuracy of the ANFIS model based on expected and observed values. Equation 9.9 shows the calculation of the RMSE values.

$$RMSE = \sqrt{\frac{\sum_{t-1}^{n} (\text{actual}(t) - \text{predicted}(t))^2}{n}} \tag{9.9}$$

where predicted (t) and actual (t) are the estimated and observed values and n is the total number of testing recordings. To assess the outcome of any conceptual framework, the ANFIS model is developing based on six essential parameters. ANFIS's work complements PLS-SEM analysis by capturing the relationships between the variables [6]. As mentioned, ANFIS is an adaptive fuzzy system that mimics human behavior by providing predictions with minimal ambiguity. ANFIS uses a fuzzy environment to mix numerical and linguistic data and processes such data using artificial neural networks. The ANFIS approach is based on a fuzzy dataset accompanying the training process, with the goal of producing accurate predictions [7, 8]. Artificial neural networks address the predictability of complex, uncertain situations. This enables the ANFIS approach to combine the properties of predicting nonlinear relationships with a fuzzy inference approach to capture nonlinear relationships and provide highly accurate predictions for the studied cases [9, 10]. Structural equation modeling suffers from its inability to capture nonlinear relationships. Therefore, the benefit of ANFIS lies in interpreting these types of relationships and making accurate predictions [11].

9.2 Previous Studies of PLS-SEM and ANFIS

The PLS-SEM and ANFIS hybrid statistical model represents one of the most modern and advanced methods. This model combines two features: explanatory and predictive. Therefore, when specialists define it, they give it a comprehensive analytical character, as it combines explaining relationships on the one hand and predicting causes on the other. In this section, we seek to shed light on studies that have adopted this modern approach, to familiarize the reader with the important role of this approach and its adoption in various disciplines. With regard to workplace safety, Çakıt et al. [3, 4] conducted an assessment of workplace safety and the extent of workers' exposure to work-related injuries. This study adopted the PLS-SEM method to test hypotheses through structural equation modeling. The study also adopted the ANFIS method to detect non-linear relationships affecting workplace safety. Therefore, the study concluded that there is a relationship between non-compliance with safety regulations and workplace hazards. Referring to traditional studies, we find that they rely on examining linear relationships using PLS-SEM, which weakens the model's ability to predict causes outside the pre-designed study model. Therefore, combining these two approaches provides greater power to capture non-linear relationships. The study revealed the importance of senior management's commitment and strict procedures for employee compliance with safety measures. It also demonstrated that there are environmental factors that influence employees and encourage them to adhere to safety procedures. In the field of hotel services, Yadegaridehkordi et al. [12] conducted a study on the elements of successful hotel services in some hotels in Malaysia. The researchers adopted perspectives on the organization, environment, and technology on the one hand, and people, organization, and technology on the other. The aim was to identify the elements that lead to the development of hotel services. The study adopted the PLS-SEM and ANFIS approaches to analyze direct and indirect relationships. The study was distinguished by its ability to capture the complex aspects of interactions between variables. The study demonstrated that customer satisfaction is one of the most important and decisive elements in the continued development of hotel services. In another industry, Ho and Tsai [13] evaluated the potential for new product development. The study focused on performance quality associated with value innovation and process quality. The same hybrid approach was used to capture complex interactions. Therefore, the study successfully captured these interactions. Khaw et al. [9, 10] investigated the sustainability potential of small and medium-sized enterprises (SMEs). They focused on elements such as top management commitment, resource availability, organizational culture, and stakeholder engagement. The study found that there are nonlinear interactions that policymakers need to address to better understand sustainability. The study focused on stakeholder engagement and top management commitment as two strong predictors. In another context, Espahbod et al. [7, 8] investigated the potential of blockchain in supply chain sustainability. The study adopted PLS-SEM and ANFIS approaches to investigate the elements of lean manufacturing and sustainability. This hybrid approach helped capture the nonlinear relationship between blockchain and supply chain sustainability. On the other hand, Alrashed et al.'s [14] SEM and ANFIS study successfully revealed the thermophysical properties of nanofluids, especially at changing temperatures. The study demonstrated the significant success of this model in predicting other factors that may lead to the same results, providing crucial predictive insights for future industrial applications. In another contribution to the field of engineering education, the factors influencing the use of metaverse in engineering education were analyzed. Wiangkham and Vongvit [6] achieved accurate results by adopting a PLS-SEM and ANFIS approach. The results revealed that technology acceptance significantly contributes to the use of metaverse technologies. The study also confirmed the existence of complex nonlinear relationships between the studied variables.

Table 9.1 includes several empirical studies that vary in terms of the nature of their variables. These variables are classified as independent, mediating, moderating, and dependent variables, depending on their role in each study. For example, in the hospitality industry, variables such as benchmarking and information technology played an influential role (independent variables) influencing financial performance outcomes (dependent variables). In terms of mediation, service quality and senior management support were pivotal factors mediating these relationships. In terms of moderation roles, leadership commitment and employee empowerment played an moderative role in strengthening or weakening the relationship between the variables. In the context of product development and sustainability, new product development, value innovation, and process quality served as independent variables. Mediating variables were organizational culture and stakeholder engagement, which led to outcomes such as sustainability commitment and performance. In the context of supply chain and blockchain innovation, the table highlights the role of supply chain analytics and LARG practices in improving sustainable performance. Here, supply chain innovation a mediating variable. In metaverse education, variables such as performance expectations, effort expectations, and social influence are influential factors in metaverse adoption intention. Technological readiness is a mediating factor that influences adoption behavior. For autonomous vehicles, perceived usefulness, trust, and safety are key factors influencing attitudes and behaviors. Social norms also contribute to user acceptance. In the context of workplace safety, management commitment, risk awareness, and training contribute to developing a positive safety culture, reducing the likelihood of accidents and increasing the sense of occupational safety. Finally, in engineering, particularly with regard to nanofluids, properties such as temperature and nanoparticle concentration affect thermal properties such as viscosity, density, and conductivity.

Table 9.1 Hybrid SEM–ANFIS application

Study theme	Independent variables	Mediators	Moderators	Dependent Variables
Hotel services sector	IT usage Financial performance Benchmarking Customer satisfaction top management support	Service quality Employee empowerment	–	Hotel success
New product development	Value innovation Process quality	–	Market uncertainty	Product development performance
SMEs sustainability	Leadership commitment Organizational culture Resource availability	Stakeholder engagement	–	Sustainability commitment
Blockchain supply chain	Blockchain analytics supply chain innovation	LARG practices (lean, agile, resilient, green)	–	Sustainable Performance
Metaverse education	Performance expectancy Effort expectancy Social influence	–	Technological readiness	Intention to adopt metaverse
Autonomous vehicles	Perceived usefulness Trust Perceived safety	Attitude	Social norms	Intention to use autonomous vehicles
Safety culture/ Workplace	Management commitment Training Risk awareness	–	–	Safety culture Perceived safety Accident likelihood
Nanofluids (Engineering)	Temperature Nanoparticle concentration	–	–	Viscosity Density Thermal conductivity

9.3 Contemporary Overview of FsQCA Analysis

This section attempts to review the role of fuzzy set qualitative comparative analysis (fsQCA) in analyzing phenomena related to the social sciences. This section focuses on deconstructing reviews conducted in the fields of entrepreneurship, marketing, tourism, and others. The review presented by Kumar et al. [15] is one of the most comprehensive. This study compiled 1155 articles published in prestigious journals. The study relied on deconstructing the visions of thought leaders on the fsQCA method. Their study resulted in maps that illustrate the most common uses of this approach. However, there is a lack of consensus regarding transparency of calibration and justification of membership thresholds. Nikou et al. [16] arguably addressed this deficiency. They presented a review study focused on transparency of calibration and proposed some theoretical solutions by referring to the best-used protocols. However, they did not provide empirical solutions in this field. Therefore, these treatments remain theoretical from the perspective of many. In the field of marketing, Gligor et al. [17] presented a study based on a compilation of studies that used the fsQCA approach in the field of marketing. They concluded that this approach is an effective tool for understanding causal complexity and asymmetry. They also recommended the importance of identifying inconsistencies in calibration logic, which jeopardizes the reliability of the results. These concerns echo the previous criticism presented by Mello [18]. Mattke et al. [19] highlighted the importance of addressing the same shortcomings. This study compiled articles related to the use of fsQCA in the field of technology. Although they highlighted the importance of this approach in understanding causal relationships, they indicated that the reliability of this approach is hampered by issues related to transparency and calibration. Kumar et al. [20] then examined the tourism sector. The study showed that fsQCA is widely used in the Northwest, particularly with regard to customer satisfaction. Therefore, they argued for the importance of focusing on other aspects such as tourism sustainability and crisis management. Geremew et al. [21] developed a PRISMA-based methodology for analyzing the use of fsQCA in hypothesis development. This emphasized the importance of exploring models other than the theory of planned behavior and the motivated intentions model. In the hospitality industry, Eluwole et al.'s [22] study criticized fsQCA's reliance on cross-sectional data only. The researchers conducted a bibliometric study focusing on this type of data and also criticized the lack of transparency in calibration methods. Diwanji [23] supported this criticism in his review of consumer behavior. His criticism also included the focus of most studies on consumer behavior in Eastern markets. Other studies by Ceglia et al. [24] and Garti et al. [25] examined fsQCA in relation to green innovation and operations management. While acknowledging the power of fsQCA in identifying configural patterns underlying good performance, they expressed significant concerns about the overuse of cross-sectional designs and inappropriate calibration practices. Therefore, Ceglia et al. advocate combining fsQCA with structural equation modeling (SEM) and neural network calibration techniques to create a model that enhances predictive capabilities. fsQCA appears to be superior to traditional statistical methods in its ability to address constitutive causality, causal asymmetry, and eventual equivalence, making it a powerful tool for analyzing complex phenomena. However, there are recurring methodological shortcomings. Chief among these is a lack of transparency, especially regarding calibration thresholds and consistency cutoffs, which reduces replicability. Additionally, many studies rely on small sample sizes (typically between 15 ane 20 cases) and cross-sectional data, which limits generalizability. Another noted weakness is the narrow geographical focus of fsQCA research. Most applications are concentrated in developed countries, particularly in industries such as tourism and marketing, with less attention paid to developing nations. Theoretical diversity is also lacking, as most studies draw primarily from the theory of planned behavior or motivational theories, with limited exploration of alternative frameworks. Table 9.2 presents eleven review studies on fsQCA, most of which focus on hospitality, marketing, and tourism. These studies employ various review methods, including systematic reviews and bibliometric analyses. The table includes essential information such as research domains, contributions, limitations, and recommendations. It serves as a useful tool for readers to quickly grasp the nature and focus of fsQCA-based studies and helps researchers identify gaps in the literature using high-quality, recent sources. Moreover, the table reveals methodological intersections between the studies, allowing for the development of applicable and impactful future research directions.

9.4 Conclusion

Based on the studies examined and analyzed, we can conclude several points. First, the fsQCA technique is steadily gaining popularity and expansion across various fields, particularly in the social sciences and entrepreneurship. fsQCA deals with multiple, complex variables, which typically require a thorough understanding and in-depth analysis. Traditional analytical models cannot perform the tasks performed by fsQCA. However, there are concerns arising from the lack of clarity in data calibration. Many studies do not clearly explain the calibration of their data. Furthermore, they sometimes do not provide adequate justifications. Therefore, the results cannot be trusted and adopted in the future. In this context, both Gligor et al. [17] and Mello [18] pointed out that using abbreviations such as percentages or providing insufficient details about calibration leads to weak or unreliable results. In addition, some studies rely on cross-sectional data rather than time series and frequency of changes. This leads to a vague view of future changes. As conditions and activities change, the data become unbalanced. Although some suggest using longitudinal data, this has not yet become common practice. There is also the question of where this method is used. Most fsQCA research appears to come from wealthier countries and focuses on consumer or business topics, neglecting public sectors and issues in developing regions.

Table 9.2 Overview of fsQCA studies by some criteria

Author(s)	Type	Study area	Contribution	Limitations	Recommendations
[15]	Methodological review	Business and management	Mapping understanding sets	Limited methodological details	More in-depth methodological review and translucency
[20]	Systematic review	Tourism and hospitality	Thematic investigation in hospitality fsQCA	Overemphasis on Global North, narrow thematic multiplicity	Extend geographic and thematic scope
[17]	Systematic review	e-Marketing	Gaps and future plans	Lack of outer validity and replica	Precise calibration protocols and replication investigations
[16]	Systematic review	Entrepreneurial orientation	Best techniques guide for fsQCA in entrepreneurial orientation	Lack of empirical testing of most useful techniques	Testing suggested improvements in practical work
[21]	Systematic review	Tourism and travel	Benefits of fsQCA as a mixed method	Limited theoretical variety, mainly planned behavior theory	Widen theoretical views
[22]	Bibliometric analysis	Hospitality and tourism	Perception of research trends	Dependence on cross-sectional data	Encourage longitudinal data and translucent calibration
[23]	Systematic review	Consumer research	Mapping configurational complexity	Rich use of secondary datasets, Western market bias	Integrate primary data and diversify geographic contexts
[19]	Methodological review	Information systems	Calls for clarity and double	Lack of publicly obtainable calibration datasets	Normalize calibration reporting, and available data sharing
[25]	Systematic review	Operations management	Identifies configurational designs	Lack of longitudinal measures	Assume longitudinal designs and diverse ways

(continued)

Table 9.2 (continued)

Author(s)	Type	Study area	Contribution	Limitations	Recommendations
[24]	Systematic review	Risk management	Combines stakeholder view with fsQCA	Little sensitivity analysis	Highlight sensitivity testing and diverse-method methods
[18]	Methodological review	Cross-disciplinary review	Exact methodological analysis	Older review, smaller focus on recent advancements	Buzzes for strict standards and translucency

References

1. Mohd Adnan, M.R.H., Sarkheyli, A., Mohd Zain, A., Haron, H.: Fuzzy logic for modeling machining process: a review. Artif. Intell. Rev. **43**(1), 345–379 (2015)
2. Sayeh, K.F., Tamalouzt, S., Ziane, D., Bekhiti, A., Belkhier, Y.: Utilizing fuzzy logic control and neural networks based on artificial intelligence techniques to improve power quality in doubly fed induction generator-based wind turbine system. Int. J. Energy Res. **2025**(1), 1–21 (2025)
3. Çakıt, E., Olak, A.J., Karwowski, W., Marek, T., Hejduk, I., Taiar, R.: Assessing safety at work using an adaptive neuro-fuzzy inference system (ANFIS) approach aided by partial least squares structural equation modeling (PLS-SEM). Int. J. Ind. Ergon. **76**(3), 1–11 (2020)
4. Çakıt, E., Olak, A.J., Karwowski, W., Marek, T., Hejduk, I., Taiar, R.: Assessing safety at work using an adaptive neuro-fuzzy inference system (ANFIS) approach aided by partial least squares structural equation modeling (PLS-SEM). Int. J. Ind. Ergon. **76**, 102925 (2020)
5. Foroughi, B., Nhan, P.V., Iranmanesh, M., Ghobakhloo, M., Nilashi, M., Yadegaridehkordi, E.: Determinants of intention to use autonomous vehicles: findings from PLS-SEM and ANFIS. J. Retail. Consum. Serv. **70**(1), 1–14 (2023)
6. Wiangkham, A., Vongvit, R.: Exploring the drivers for the adoption of metaverse technology in engineering education using PLS-SEM and ANFIS. Educ. Inf. Technol. **29**(6), 7385–7412 (2024)
7. Espahbod, S., Tashakkori, A., Mohsenibeigzadeh, M., Zarei, M., Arani, G.G., Dzikuć, M., Dzikuć, M.: Blockchain-driven supply chain analytics and sustainable performance: analysis using PLS-SEM and ANFIS. Sustainability **16**(15), 1–27 (2024)
8. Espahbod, S., Tashakkori, A., Mohsenibeigzadeh, M., Zarei, M., Arani, G.G., Dzikuć, M., Dzikuć, M.: Blockchain-driven supply chain analytics and sustainable performance: analysis using PLS-SEM and ANFIS. Sustainability **16**(15), 6469 (2024)
9. Khaw, K.W., Sadaa, A.M., Alnoor, A., Zaidan, A.S., Ganesan, Y., Chew, X.: Spurring sustainability commitment strategy of family-owned SMEs: a multi-analytical SEM & ANFIS perspective. J. High Technol. Managem. Res. **34**(1), 1–16 (2023)
10. Khaw, K.W., Sadaa, A.M., Alnoor, A., Zaidan, A.S., Ganesan, Y., Chew, X.: Spurring sustainability commitment strategy of family-owned SMEs: a multi-analytical SEM & ANFIS perspective. J. High Technol. Managem. Res. **34**(1), 100453 (2023)
11. Baig, M.I., Yadegaridehkordi, E., Shuib, L., Sallehuddin, H.: Identifying determinants of big data adoption in the higher education sector using a multi-analytical SEM-ANN approach. Educ. Inf. Technol. **28**(12), 16457–16484 (2023)
12. Yadegaridehkordi, E., Nilashi, M., Nasir, M.H.N.B.M., Ibrahim, O.: Predicting determinants of hotel success and development using Structural Equation Modelling (SEM)-ANFIS method. Tour. Manage. **66**, 364–386 (2018)
13. Ho, Y.C., Tsai, C.T.: Comparing ANFIS and SEM in linear and nonlinear forecasting of new product development performance. Expert Syst. Appl. **38**(6), 6498–6507 (2011)
14. Alrashed, A.A., Gharibdousti, M.S., Goodarzi, M., de Oliveira, L.R., Safaei, M.R., Bandarra Filho, E.P.: Effects on thermophysical properties of carbon based nanofluids: experimental data, modelling using regression, ANFIS and ANN. Int. J. Heat Mass Transf. **125**, 920–932 (2018)
15. Kumar, S., Sahoo, S., Lim, W.M., Kraus, S., Bamel, U.: Fuzzy-set qualitative comparative analysis (fsQCA) in business and management research: a contemporary overview. Technol. Forecast. Soc. Chang. **178**, 121599 (2022)
16. Nikou, S., Mezei, J., Liguori, E.W., El Tarabishy, A.: FsQCA in entrepreneurship research: opportunities and best practices. J. Small Bus. Manage. **62**(3), 1531–1548 (2024)
17. Gligor, D.M., Golgeci, I., Rego, C., Russo, I., Bozkurt, S., Pohlen, T., et al.: Examining the use of fsQCA in B2B marketing research: benefits, current state and agenda for future research. J. Bus. Ind. Mark. **37**(7), 1542–1552 (2022)
18. Mello, P.A.: From prospect to practice: a critical review of applications in fuzzy-set qualitative comparative analysis. (2013). Available at SSRN 2349513

19. Mattke, J., Maier, C., Weitzel, T., Thatcher, J.B.: Qualitative comparative analysis in the information systems discipline: a literature review and methodological recommendations. Internet Res. **31**(5), 1493–1517 (2021)
20. Kumar, S., Sahoo, S., Ali, F., Cobanoglu, C.: Rise of fsQCA in tourism and hospitality research: a systematic literature review. Int. J. Contemp. Hosp. Manag. **36**(7), 2165–2193 (2024)
21. Geremew, Y.M., Huang, W.J., Hung, K.: Fuzzy-set qualitative comparative analysis as a mixed-method and analysis technique: a comprehensive systematic review. J. Travel Res. **63**(1), 3–26 (2024)
22. Eluwole, K.K., Lasisi, T.T., Parvez, M.O., Cobanoglu, C.: Application of fuzzy-set qualitative comparative analysis (fsQCA) in hospitality and tourism research: a bibliometric study. J. Hospitality Tourism Insights **7**(5), 3032–3054 (2024)
23. Diwanji, V.S.: Fuzzy-set qualitative comparative analysis in consumer research: a systematic literature review. Int. J. Consum. Stud. **47**(6), 2767–2789 (2023)
24. Ceglia, I., Battaglia, M., Iandolo, F., Calabrese, M.: A systemic view of risk management and the role of stakeholders. Literature review and fsQCA analysis. In Euram Conference, pp. 1–40. (2023)
25. Garti, M.O., Arif, J., Jawab, F.: Application of fsQCA method in operations management area: systematic literature review. In: Intersecting Entrepreneurship, Internationalization, and Green Innovation, pp. 291–322. (2025)

Part IV
PLS-SEM and MCDM Methods

Chapter 10
Multi-criteria Decision-Making

Abstract This chapter reviews multi-criteria decision-making techniques and divides them into two main categories: criterion weighting and alternative ranking. These techniques are reliable tools for addressing complex problems and provide a structured and transparent framework that takes multiple criteria into account. The chapter explains how these tools can be used in business and management, and discusses the potential for combining them with fuzzy methods and structural equation modeling to enhance decision-making flexibility. Finally, practical applications reflect the importance of these methods in business and management, as they support more balanced and accurate decision-making.

Learning Objectives

- Decision-making methods are divided into two types: determining criteria weights and ranking alternatives.
- MCDM methods are important in supporting organizational decision-making; they provide a structured and transparent process that considers multiple criteria, helping achieve more balanced and objective choices.
- MCDM methods can be combined with fuzzy methods and structural equation modeling to better handle uncertainty.
- MCDM methods have many applications in the field of business and management.

10.1 Decision-Making Process

It is a fact that people make decisions now and then. Decision-making is one of the most significant and quite challenging processes in the management of any organization. No decision can be made without going through the decision-making process [1]. Decision-making can be described as a problem-solving process that has to identify the desirable outcomes. This process can be rational or irrational, implicit or explicit, and influenced by many factors, such as physiological, cultural, and social

X. Chew et al., *Partial Least Squares Structural Equation Modeling and Complementary Methods in Business Research*, Information Systems Engineering and Management 67, https://doi.org/10.1007/978-3-032-01055-1_10

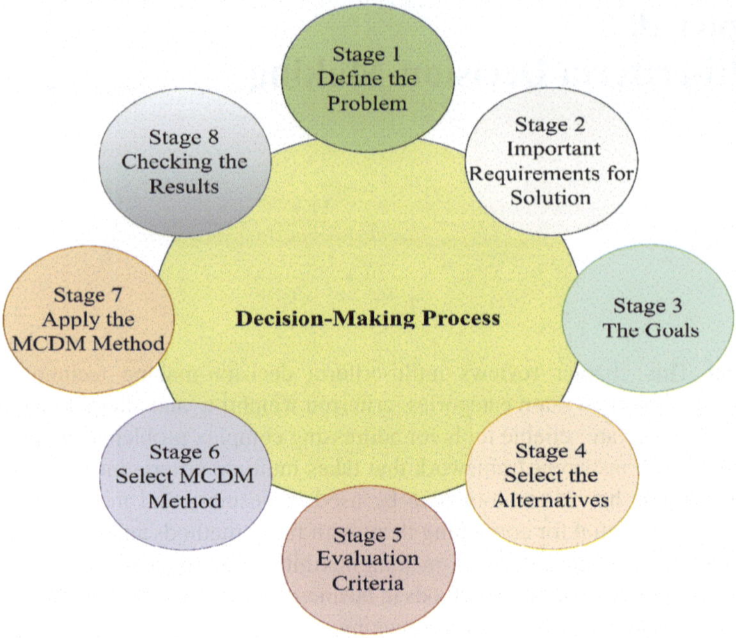

Fig. 10.1 Decision making process

[2]. All these aspects can determine the difficulty level of this process. In the present generation, any decision-making problem is solved by using mathematics, statistics, manpower, economics theories, and computers to estimate the solution to the decision-making problems [3]. Figure 10.1 shows the steps of decision making.

As shown in Fig. 10.1, the most common approach to decision-making process is as follows:

(1) Define the problem. (2) Important requirements for solution. (3) The goals that need achievement. (4) Select the alternatives of solutions in order to get the best one. (5) proper evaluation criteria. (6) Select the right multi-criteria decision-making method to find the key to the issues. (7) Apply the MCDM method to the decision matrix established from the intersection of criteria and alternatives. (8) Checking the results by sensitivity analysis test or other testes.

10.2 Multi-criteria Decision-Making

MCDA is one of the most reliable techniques for decision making and it is said to be a revolution in this field [4]. The first research study on the multi-criteria decision-making was done by Benjamin Franklin when he published the concept of moral algebra. Starting from the 1950s various empirical and theoretical scientists have tried to establish MCDM methods in order to assess the mathematical modeling effectiveness of these methods that could assist in structuring the decision-making process and preferences for the alternatives. In literature [5–7], many terms have been used in referring to MCDM and they are as shown below:

- Multi-Criteria Decision Analysis (MCDA)
- Multi-Objective Decision Making (MODM)
- Multi-Attributes Decision Making (MADM)
- Multi-Dimensions Decision-Making (MDDM).

10.3 MCDM Methods

Literature review reveals that there are several types of MCDDM methods which have been developed or enhanced by various authors in the last few decades. The key distinctions between these methods are based on the level of algorithm complexity, criteria weighting techniques, preference evaluation criteria depiction, uncertainty handling, and data aggregation technique [8–10]. Figure 10.2 shows the maximum frequently MCDM methods in businesses filed.

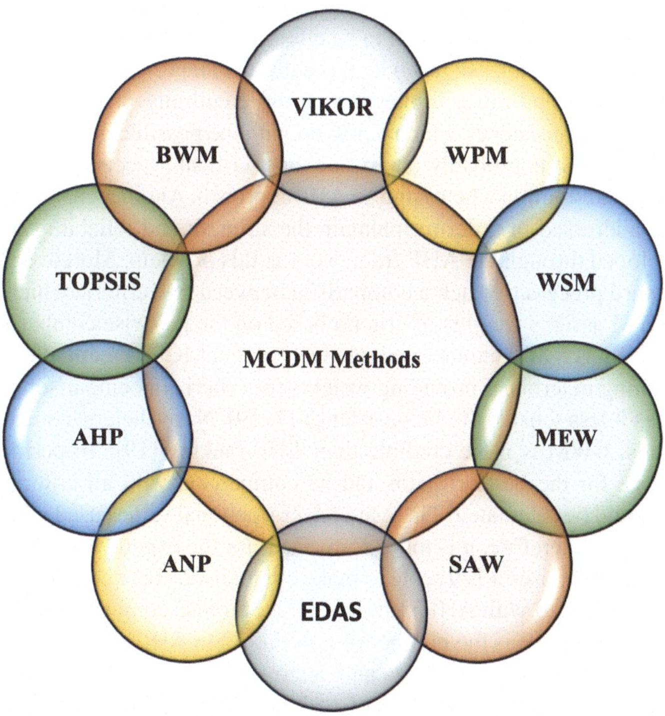

Fig. 10.2 MCDM methods

Generally, MCDM methods can be categorised into two major types namely the mathematical and the human [11]. The mathematical assessment approach uses the information on each factor to determine the factor's weights. Some of the most effective methods for the purpose include CRITIC [5], VIKOR [12], TOPSIS [13], MEW [14]. These methods are advantageous over the others as they do not involve subjective decision making in the determination of criterion significance. Every criterion is assigned a weight in accordance with some equation. These techniques have been used in the literature review and do not cause any contradictions; however, the subjective weights of the findings still remain the key factors that affect the final ranking, as they reflect the opinions of experienced professionals [15]. Weights that are obtained for the assessment criteria may be fine-tuned according to changes which have been made to the row data. Therefore, the process of assigning weights in these methods is imprecise [16]. Such procedures cannot capture the knowledge and subjective value of the decision-makers for instance if the criteria must be weighted subjectively.

The human assessment gives importance to the attributes which are based on the knowledge and skills of the decision makers. Human weights show the wisdom of the decision makers and the feedbacks from previous experiences. The Analytic Hierarchy Process (AHP), the Best–Worst Analysis (BWM), and Analytic Network Process (ANP) are some examples of human evaluation methods as illustrated below. Since MCDM is dependent on the opinions of an expert and his or her decision making, AHP has been used in many fields. This technique was put forward by Saaty [16]. In spite of this, AHP is deemed to be effective; one of the major drawbacks of the AHP approach is inconsistency which is a disadvantage. This inherent limitation is a major problem that raises a question on the credibility of the AHP approach [17]. The issue of consistency which has to do with the possible inconsistency in the pairwise comparison is probably one of the most critical aspects that needs proper consideration and solution when using an AHP approach. Another challenge that also needs to be addressed in order to maintain the strength and reliability of decision results developed through the AHP framework is this problem. Moreover, it is not a straightforward process to reach a comparison between two criteria which are not.

Also, BWM assigns weights of criteria based on the pairwise comparison of the criteria sets [18]. BWM is another and relatively recent MCDM method which is less time-consuming in terms of providing weights for criteria in comparison with AHP, which makes it less sensitive to inconsistency [7, 19]. Nevertheless, comparing with both methods, BWM is more credible than AHP ranking [19]. To perform BWM the preference for the best criterion and its comparison with all criteria over the worst one should be calculated. There are fewer fractional values used while making comparisons and, therefore, it is more convenient for the experts than other MCDM methods. The BWM uses a 1–9 scale to make comparisons between two items in the model [20]. Compared with AHP's $(n\,(n-1)/2)$ pairwise comparisons, BWM only needs $(2n-3)$, where n is the number of criteria. As in AHP, this approach has the same problems in ranking the criteria and determining the most and least important criteria as well as the relative weights of each criterion [7]. When comparing items on a scale basis, for example from 1 to 9, the DM has to state how many times one criteria was better than the other on that scale. By virtue of the subjective comparisons, the comparisons made here were not normal, thus, more mental effort was needed to make the comparison. Furthermore, the original and the extended BWM have presented several measures of consistency and still, all have weaknesses [21].

On the other hand, there are some specific drawbacks in some MCDM methods, such as comparison between the criteria, which are not in any way related or even comparable. Some MCDM models assume that all criteria can be compared with each other without any problem. This approach brings several problems with it. There are criteria such as sound and color that are discernible by different senses, namely the auditory and the visual [22, 23]. Comparing them within the same framework may give wrong results since there is no way of measuring both to the same certain yardstick. In comparison with the unrelated criteria, the assigning the weights is subjective. It is quite possible that decision-makers may not be able to give the right weightage to the criteria, which are not measurable, and hence the decision-making may not be very accurate [24, 25]. It is possible to make some criteria underrated or overrated based on personal preferences. Moreover, most of the MCDM techniques operate on the basis of distance in order to find out how close or far an alternative is from the ideal or anti-ideal solution. Usually, these methods suppose that improving or worsening a criterion's value is always beneficial. However, this assumption can be dangerous, particularly when working with criteria for which neither the maximum nor minimum is desirable, such as many examples in accounting filed. Finally, another significant issue that is also observed in the MCDM methods (i.e., human and mathematical approaches) are imprecise and uncertainty information. Experts work with linguistic constructs and cannot define the weights in real numbers. Therefore, these issues, including this information are not easy to address [26].

10.3.1 Benefits Related to Using MCDM

The focus on MCDM has been rising since it is seen as a crucial component of operation research, decision science, and expert systems that involve multiple alternatives and criteria in a business setting [27]. The primary purpose of MCDM is to identify the most suitable option among the possible options based on the given criteria [28]. MCDM techniques can provide the solutions in many areas of knowledge like management, society, economy, business and engineering and other.

Most of the MCDM approaches employ subjective or objective weight since the values of the alternatives are different concerning their ranking on the basis of their significance. Numerous researches have been conducted regarding the importance of assessing utterly unfamiliar or interpretable indexes [29]. Here, in real life, we can hardly ever encounter a condition with full or partial knowledge. It can be observed that the MADM problem has to be solved using the fuzzy set theory in most instances.

Currently, for instance, TOPSIS is a very popular and well-known approach for solving MCDM problems [30–32], but not a suitable approach for all MCDM problems. In practical problems, the researchers often try to choose as many of the existing techniques as possible that would best suit the experts' preferences and the situation at hand. In general, techniques to solve a problem in MCDM are context-dependent problems.

10.3.2 Benefits Related to Using Fuzzy Set with MCDM

In MCDM issues sometimes the choosing attributes are based on the expert's judgment. The reason is that the selection is made on the basis of data or personal experience. Human thoughts are vague, which in turn results in wrong and partial decisions. The fuzzy MCDM techniques are quite suitable in the decision-making process of evaluating existing alternatives in order to choose the best one when the criteria used have the notion of subjective perception [33]. Thus, it is concluded that the evaluation process must be operating under a fuzzy environment in order to allow linguistic variables. The problem with this method is that it is not very certain and is quite subjective, which can lead to wrong weights and challenges in selecting the criterion weights [34]. For instance, in the accounting field selection problem, the decision-making process is affected by uncertainty as imprecision is inherent in the assessment of intangible factors [35]. Different DMs also set the level of importance that leads to imprecise weight of the criteria. The fuzzy set theory is one of the most effective techniques used to address uncertainty in accounting and business issues; the selection problem, in general, many authors work in a group MCDM environment where there is a lot of ambiguity and imprecision and deal with qualitative or quantitative criteria and different stakeholders' viewpoints [36, 37]. Therefore, the concept of a linguistic variable is quite helpful in dealing with situations where cases are somewhat more complicated or not strictly defined to be reasonably quantified in any conventional quantitative manner, and fuzzy numbers are used to effectively capture linguistic variables [38, 39]) was the first to use fuzzy set theory with MCDM problems which are identified by the subjective judgment as an efficient way of solving the problems of vagueness, lack of information and ambiguity in human decision making process which is known as FMCDM [40, 41]. Thus, the fuzzy set theory is a powerful tool in depicting a condition of uncertainty or another type of a fuzzy environment that is typical of the area and behavior of managerial decisions in the [19, 42]. The most important contribution of the fuzzy set theory is the possibility of applying mathematical operators in the fuzzy environment by means of processing the imprecise data. However, the analysts may sometimes face problems in the assessment of the weights of criteria as numbers in interval [0,1] and the membership degree of each element of the fuzzy sets used to rank the alternatives with regard to criteria and/or the values of criterion weights. Hence, the IVFS-MCDM approaches need to be used when dealing with fuzzy MCDM issues [43], trapezoidal membership function [44] or triangular fuzzy numbers [45]. The authors were able to use fuzzy set in many fields of research in order to address the issue of uncertainty and vagueness. Moreover, the information about DMs is not fully understood and described, thus, it is rather imprecise. To overcome this situation (uncertainty and hesitation), IFSs are employed to choose the best DM preferences. IFS is a robust approach because it handles membership and non-membership functions and also [46]. In practical problems one can meet the lack of information or the information which is impossible to obtain. For instance, it is not easy to assign real numbers to evaluate the level of uncertainty in human judgment preferences. It is very frequently the case that the information is not well-defined and therefore, vagueness and fuzzy sets are ubiquitous. Due to the fact that the description of linguistic information is qualitative, some of the criteria for evaluation and the alternatives are relative. Thus, to avoid the above issues, the researchers opted for fuzzy logic [44, 46].

10.3.3 Benefits Related to Using MCDM with Business Applications

It is also important to note that MCDM plays a vital role in the field of business in all its forms, including trading, financial markets, firms, and suppliers in the management of commercial markets [5, 29]. Companies need to create new products in order to sustain their value in market competition, especially in the global market. Therefore, improving new product platforms and new products that satisfy reasonable demands of quality, price, safety, and performance is crucial. Several decision points are identified in each stage of the project where the management makes decisions about the project. The choice of the idea of a new product is one of the most important decisions that are made in the course any project management. Numerous research works suggest that managers should complete projects once they, for instance, green product development [47] and project management [48]. Furthermore, the population in cities is high, and the economy is also growing. Therefore, it is necessary to apply MCDM methods in order to enhance the management of the economy in large cities. Companies should also pay attention to good planning to be able to offer a large number of services. Such companies usually rely on their employees' experience. Hence, decision-making techniques can be effectively used and have become essential in heterogeneous networks [49].

Other aspects of road safety, public transport, and ports, and to make the best out of them, it has to be done in a scientific manner. One of the techniques applied to handle and achieve the best solution to the problems and assess those services is MCDM. MCDM are also applied in companies to control and create frameworks that ensure the success of the company [50]. Some of the applications include electric power management, risk management, and safety management [51].

MCDM in business has the following benefits due to the fact that it offers a systematic way of solving decision-making problems, which may involve several criteria, some of which are opposing. It enhances decision-making by taking into account various factors, including cost, risk, performance, and customer satisfaction, and this makes the outcome of the decision to be sound. MCDM improves the decision-making processes by increasing the level of openness and involving the stakeholders in the decision-making process, thus eliminating biases. Through the use of both quantitative and qualitative data, the organization can solve problems, minimize risks, and make strategic decisions that are in line with the organization's objectives, thus enhancing efficiency and productivity Karbassi Yazdi et al. [34, 50, 52].

10.3.4 Benefits Related to Using Structural Equation Modeling with MCDM

The combination of Structural Equation Modeling (SEM) and MCDM is a very significant methodological approach in solving various decision-making problems especially when the aim is to assess several criteria and their effects on a number of options [53]. Although both SEM and MCDM are very useful methodologies on their own, the synergy of both methods leads to a more effective decision-making process. This integration helps the researchers not only to determine the interconnection of variables but also the priority of the criteria and alternatives. The major advantage of integrating SEM with MCDM is that these two techniques can be quite synergistic, since none of them has all the shortcomings of the other [53, 54].

First, SEM is mostly applied in studying the link between the theoretical constructs, the hidden or unmeasured variables, and variables that are measured. This is especially useful in the empirical testing of theoretical models, where the nature of the relationships between the variables is proposed on the basis of theory or prior research [55]. SEM enables the researcher to examine the effect of several independent variables on the dependent variable at the same time; thus, it is an efficient method for analyzing the structure of data and hypotheses. However, there is a disadvantage of SEM in that although it does well in establishing relationships between variables, it does not offer a way of ordering criteria or alternatives [56]. Relative importance of the criteria cannot be determined with SEM, and the alternatives cannot be ranked in a decision analysis [53, 57]. This is a major deficiency in a situation where decision takers require guidance on which of the criteria to adopt as well as on the tradeoff between the options.

On the other hand, there are MCDM techniques like AHP, CoCoS or Fuzzy MCDM techniques which are specifically developed to evaluate criteria and rank the alternatives on the basis of those criteria [26]. MCDM offer a clear methodology of decision making by weighting certain criteria in order to recognize the most significant factors playing a role in the decision-making process. Furthermore, MCDM techniques assistance in ranking the options which is helpful for the decision makers to opt for the best one among the options provided with the weighted criteria [58, 59]. However, MCDM is good at criteria weighting and alternative ranking, which is not an adequate approach when dealing with the relationships between variables that are characteristic of SEM [60]. However, it is a weakness of MCDM that it does not consider the causal relations between the variables which are well captured by SEM [60].

Thus, the above analysis shows that the synergy of SEM and MCDM is in their integration. This paper has shown that the integration of SEM to confirm the structure of the variables and MCDM to assign weights and rank the alternatives can provide decision makers with valuable information more than the use of one approach. In particular, during the first stage, it is possible to check whether the relationships between the latent variables and their indicators are correct, which means that the theoretical model is valid and properly reflects the real world [61]. After confirming the relationships between the variables using SEM, it is then possible to perform MCDM to assess the relative weights of the criteria as well as to rank the alternatives with respect to these criteria.

The advantage of integrating SEM with MCDM is most relevant in situations wherein decision makers are faced with difficult problems that are characterized by factors and interconnections on one hand and ranking of options on the other. For instance, in the context of environmental management, SEM can be employed to depict the interdependencies of environmental, social and economic factors and MCDM may then be employed to evaluate the different policy options with regard to these factors [62]. Likewise, in business strategy, SEM can be applied for analysing the market conditions, competitive forces and organisational resources and capabilities; MCDM can be applied to rank the strategic options based on these factors. Additionally, in the context of social commerce platforms in rural communities, authors used SEM and MCDM methods to measure the trustworthiness of costumers [63].

One more advantage is that SEM can be combined with MCDM methods for consideration of both quantitative and qualitative factors in a decision-making process. SEM is particularly appropriate for coping with the unobserved factors that are frequently categorical in character, including client satisfaction levels, corporate culture, or buyer conduct [64]. These hidden variables are not easy to assess quantitatively, but through SEM it is possible to represent them with measured variables. MCDM, however, is quite suitable in handling quantitative data especially when it comes to the assessment of weights and ranking of the alternatives [42]. Through the integration of SEM and MCDM, it is possible to engage both qualitative and quantitative data in decision making thus giving a better perspective of the decision-making process.

10.3.5 Benefits Related to Using Fuzzy Delphi Method with Business Research

There are several advantages associated with the application of the Fuzzy Delphi Method (FDM) in business research, especially in conditions of uncertainty, imprecision, and the need to incorporate the opinions of experts [65]. FDM is a variation of the Delphi technique, which is based on the filtration of opinions of the experts to reach a consensus in a systematic and sequential manner. Fuzzy logic improves the ability of FDM to deal with imprecision uncertainty and subjective assessment, and this is valuable in the business arena where incomplete and inaccurate data are common, and decisions are made based on experience [65]. Below are some key benefits of using Fuzzy Delphi in business research:

Firstly, the strength of FDM is its capacity to deal with uncertainty and ambiguity, which are apparent in business research. It can also be stated that information is scarce and uncertain in most business environments, such as in market research, consumer behavior, or strategy making. Fuzzy logic enables a mathematical approach to this imprecision through the use of degrees of truth instead of the classical truth values of true or false. In FDM, it is possible to put the opinions of the experts in words, for instance, 'highly likely' or 'somewhat important', which are then quantified as fuzzy numbers. This aspect is vital in business research because data is seldom accurate, and the judgment of an expert is usually complex [65].

Secondly, FDM is based on the conventional Delphi method, which is a systematic approach of eliciting the views of the experts and gradually converging them towards a common set of opinions. This is especially so in business research, which entails solving problems that may have many dimensions, thus needing the input of more than one specialist. For instance, in the process of new product development, the representatives of marketing, production, finance, and R&D departments will have their point of view. FDM enables the gathering and assimilation of these numerous views as well as the variability in the perception of each of the experts. Through these successive iterations of feedback, the method moves towards an agreed upon position that is less impulsive and is more holistic in its approach [66].

Thirdly, in business research, decisions are mainly about financial consequences, operational consequences, or strategic consequences. FDM enhances the decision-making process by assisting in incorporating knowledgeable inputs to decision making, especially in cases where numerical data is insufficient. FDM can thus be said to eliminate or at least minimize the effects of bias in the decision-making process by quantifying subjective judgments and making the evaluation process more than once in order to make a final decision or recommendation [67]. This minimizes the chances of wrong decision making because of inadequate or contradictory information which is a frequent phenomenon in organizations.

Fourthly, as for the multi-criteria analysis, which is widespread in business research, it is crucial to determine the priority level of the criteria. This is because FDM is able to provide a more precise way of assigning weights to criteria due to the fact that it incorporates the element of uncertainty and subjectivity that is present in the judgment of experts. For instance, when evaluating how successful a particular new product will be, factors like market potential, competitive advantage, and production feasibility may all be relevant but none of them may be easily quantifiable. FDM offers the decision maker a way of systematically allocating weights to the criteria which can be used in reaching the decision, and it ensures that all the relevant factors have been considered in the decision-making process [67, 68].

Finally, the conventional ways of obtaining expert judgment may have their own shortcomings, including the dominance of some individuals during group meetings or the prejudice that is set at the beginning and carried throughout the subsequent rounds of the assessment. FDM avoids these problems by capturing the responses in an anonymous manner and by emphasizing on the sequential improvement of opinions. Others are advised to change their opinions after getting the review of other experts in order to minimize the effects of extreme values and prejudices. This results in more impartial findings, which, in turn, makes the research results more credible [69].

References

1. Chen, L., Pan, W.: Review fuzzy multi-criteria decision-making in construction management using a network approach. Appl. Soft Comput. **102**, 107103 (2021). https://doi.org/10.1016/j.asoc.2021.107103
2. Shepherd, N.G., Rudd, J.M.: The influence of context on the strategic decision-making process: a review of the literature. Int. J. Manag. Rev. **16**(3), 340–364 (2014)
3. Bruch, E., Feinberg, F.: Decision-making processes in social contexts. Ann. Rev. Sociol. **43**(1), 207–227 (2017)
4. Thakkar, J.J.: Multi-criteria Decision Making, vol. 336. Springer (2021)
5. Ali, J., Hussain, K.N., Alnoor, A., Muhsen, Y.R., Atiyah, A.G.: Benchmarking methodology of banks based on financial sustainability using CRITIC and RAFSI techniques. Decis. Making: Appl. Manage. Eng. **7**(1), 315–341 (2024)
6. Kazimieras Zavadskas, E., Antucheviciene, J., Kar, S.: Multi-objective and multi-attribute optimization for sustainable development decision aiding. Sustainability **11**(11), 3069 (2019)
7. Muhsen, Y.R., Zubaidi, S.L., Husin, N.A., Alnoor, A., Božanić, D., Hashim, K.S.: The weight fuzzy judgment method for the benchmarking sustainability of oil companies. Appl. Soft Comput. **161**, 111765 (2024)
8. Awajan, I., Mohamad, M., Al-Quran, A.: Sentiment analysis technique and neutrosophic set theory for mining and ranking big data from online reviews. IEEE Access **9**, 47338–47353 (2021). https://doi.org/10.1109/ACCESS.2021.3067844
9. Paul, V.K., Chakraborty, S., Chakraborty, S.: An integrated IRN-BWM-EDAS method for supplier selection in a textile industry. Deci. Mak.: Appl. Manage. Eng. **5**(2), 219–240 (2022)
10. Torkayesh, A.E., Deveci, M., Karagoz, S., Antucheviciene, J.: A state-of-the-art survey of evaluation based on distance from average solution (EDAS): developments and applications. Expert Syst. Appl. **221**, 119724 (2023)
11. Muhsen, Y.R., Husin, N.A., Zolkepli, M.B., Manshor, N., Al-Hchaimi, A.A.J., Ridha, H.M.: Enhancing NoC-Based MPSoC performance: a predictive approach with ANN and guaranteed convergence arithmetic optimization algorithm. IEEE Access (2023)
12. Krishnan, E., Mohammed, R., Alnoor, A., Albahri, O.S., Zaidan, A.A., Alsattar, H., Albahri, A.S., Zaidan, B.B., Kou, G., Hamid, R.A., Alamoodi, A.H., Alazab, M.: Interval type 2 trapezoidal-fuzzy weighted with zero inconsistency combined with VIKOR for evaluating smart e-tourism applications. Int. J. Intell. Syst. **36**(9), 4723–4774 (2021). https://doi.org/10.1002/int.22489
13. Khatari, M., Zaidan, A.A., Zaidan, B.B., Albahri, O.S., Alsalem, M.A., Albahri, A.S.: Multidimensional benchmarking framework for AQMs of network congestion control based on AHP and Group-TOPSIS. Int. J. Inf. Technol. Decis. Mak. **20**(05), 1409–1446 (2021)
14. Jia, F., Liu, Y., Wang, X.: An extended MABAC method for multi-criteria group decision making based on intuitionistic fuzzy rough numbers. Expert Syst. Appl. **127**, 241–255 (2019)
15. Zafar, S., Alamgir, Z., Rehman, M.H.: An effective blockchain evaluation system based on entropy-CRITIC weight method and MCDM techniques. Peer-to-Peer Netw. Appl. **14**(5), 3110–3123 (2021)
16. Zhu, G.-N., Hu, J., Ren, H.: A fuzzy rough number-based AHP-TOPSIS for design concept evaluation under uncertain environments. Appl. Soft Comput. **91**, 106228 (2020)

17. Munier, N., Hontoria, E.: Uses and Limitations of the AHP Method. Springer (2021)
18. Ajrina, A.S., Sarno, R., Ginardi, R.V.H.: Comparison of AHP and BWM methods based on geographic information system for determining potential zone of Pasir Batu mining. In: 2018 International Seminar on Application for Technology of Information and Communication, pp. 453–457 (2018).https://doi.org/10.1109/ISEMANTIC.2018.8549818
19. Ahmed, A.D., Salih, M.M., Muhsen, Y.R.: Opinion weight criteria method (OWCM): a new method for weighting criteria with zero inconsistency. IEEE Access (2024)
20. Corrente, S., Greco, S., Rezaei, J.: Better decisions with less cognitive load: the Parsimonious BWM. Omega **126**, 103075 (2024)
21. Haseli, G., Sheikh, R., Wang, J., Tomaskova, H., Tirkolaee, E.B.: A novel approach for group decision making based on the best–worst method (G-bwm): application to supply chain management. Mathematics **9**(16), 1881 (2021)
22. Baqer, N.S., Albahri, A.S., Mohammed, H.A., Zaidan, A.A., Amjed, R.A., Al-Bakry, A.M., Albahri, O.S., Alsattar, H.A., Alnoor, A., Alamoodi, A.H.: Indoor air quality pollutants predicting approach using unified labelling process-based multi-criteria decision making and machine learning techniques. Telecommun. Syst. 1–23 (2022)
23. Husin, N.A., Abdulsaeed, A.A., Muhsen, Y.R., Zaidan, A.S., Alnoor, A., Al-mawla, Z.R.: Evaluation of metaverse tools based on privacy model using fuzzy MCDM approach. In: International Multi-Disciplinary Conference-Integrated Sciences and Technologies, pp. 1–20 (2023)
24. Al Ogaili, R.R.N., Raheem, O.A., Abdkhaleq, M.H.G., Alyasseri, Z.A.A., Alsaidi, S.A.A.A., Alsaeedi, A.H., Muhsen, Y.R., Manickam, S.: AntDroidNet cybersecurity model: a hybrid integration of ant colony optimization and deep neural networks for android malware detection. Mesopotamian J. CyberSecurity **5**(1), 104–120 (2025)
25. Alamoodi, A.H., Mohammed, R.T., Albahri, O.S., Qahtan, S., Zaidan, A.A., Alsattar, H.A., Albahri, A.S., Aickelin, U., Zaidan, B.B., Baqer, M.J.: Based on neutrosophic fuzzy environment: a new development of FWZIC and FDOSM for benchmarking smart e-tourism applications. Complex Intell. Syst. 1–25 (2022)
26. Alnoor, A., Muhsen, Y.R., Husin, N.A., Chew, X., Zolkepli, M.B., Manshor, N.: Z-cloud rough fuzzy-based PIPRECIA and CoCoSo integration to assess agriculture decision support tools. Int. J. Fuzzy Syst. **27**(1), 190–203 (2025). https://doi.org/10.1007/s40815-024-01771-7
27. Kumar, R.: A Comprehensive review of MCDM methods, applications, and emerging trends. Decis. Mak. Adv. **3**(1), 185–199 (2025)
28. Chowdhury, P., Paul, S.K.: Applications of MCDM methods in research on corporate sustainability: a systematic literature review. Manage. Environ. Qual.: Int. J. **31**(2), 385–405 (2020)
29. Kazimieras Zavadskas, E., Antucheviciene, J., Chatterjee, P.: Multiple-criteria decision-making (MCDM) techniques for business processes information management. Information **10**(1), 4 (2018)
30. Geng, X., Ma, Y.: N-intuitionistic polygonal fuzzy aggregation operators and their application to multi-attribute decision making. IEEE Access **8**, 162903–162916 (2020). https://doi.org/10.1109/ACCESS.2020.3018957
31. Malakar, K., Lu, C.: Measuring sustainability as distance to ideal position of economy, society and environment: application to China's provincial water resources (2004–17). J. Environ. Manage. **292**, 112742. https://doi.org/10.1016/j.jenvman.2021.112742
32. Wang, Z., Wang, Y., Xu, G., Ren, J.: Sustainable desalination process selection: Decision support framework under hybrid information. Desalination **465**, 44–57 (2019). https://doi.org/10.1016/j.desal.2019.04.022
33. Al-Samarraay, M.S., Zaidan, A.A., Albahri, O.S., Pamucar, D., AlSattar, H.A., Alamoodi, A.H., Zaidan, B.B., Albahri, A.S.: Extension of interval-valued Pythagorean FDOSM for evaluating and benchmarking real-time SLRSs based on multidimensional criteria of hand gesture recognition and sensor glove perspectives [Formula presented]. Appl. Soft Comput. **116** (2022).https://doi.org/10.1016/j.asoc.2021.108284

34. Puška, A., Nedeljković, M., Božanić, D., Štilić, A., Muhsen, Y.R.: Evaluation of agricultural drones based on the COmpromise Ranking from Alternative SOlutions (CORASO) methodology. Eng. Rev. **44**(4-SI 2024), 77–90 (2024)
35. Elmassri, M.M., Harris, E.P., Carter, D.B.: Accounting for strategic investment decision-making under extreme uncertainty. Br. Account. Rev. **48**(2), 151–168 (2016)
36. Alhasnawi, M.Y., Al-Hchaimi, A.A.J., Muhsen, Y.R., Lekmiti, A.: A bibliometric review of trends and insights of internet of things on cybersecurity issues. Curr. Future Trends AI Appl. **1**, 127–147 (2025)
37. Hamal, S., Senvar, O.: A novel integrated AHP and MULTIMOORA method with interval-valued spherical fuzzy sets and single-valued spherical fuzzy sets to prioritize financial ratios for financial accounting fraud detection. J. Intell. Fuzzy Syst. **42**(1), 337–364 (2022)
38. Zadeh, L.A.: Fuzzy sets. Inf. Control **8**(3), 338–353 (1965). https://doi.org/10.1016/S0019-9958(65)90241-X
39. Zadeh, L.A.: A note on Z-numbers. Inf. Sci. **181**(14), 2923–2932 (2011)
40. Jumarni, R.F., Zamri, N.: A new concept of fuzzy TOPSIS and fuzzy logic in a multi-criteria decision. In: Ghazali, R., Deris, M.M., Nawi, N.M., Abawajy, J.H. (eds.) vol. 700, pp. 161–170. Springer Verlag (2018). https://doi.org/10.1007/978-3-319-72550-5_16
41. Zailani, N.A.M., Ahmad, S.A.S., Mohamad, D.: A fuzzy rule-based Fuzzy Inferior Ratio method with reliability factor. **1988** (2021). https://doi.org/10.1088/1742-6596/1988/1/012059
42. Al-Hchaim, A.A.J., Muhsen, Y.R., Gwad, W.H., Alkayal, E.S., Al Ogaili, R.R.N., Alyasseri, Z.A.A., Alnoor, A.: Prioritizing network-on-chip routers for countermeasure techniques against flooding denial-of-service attacks: a fuzzy multi-criteria decision-making approach. Comput. Model. Eng. Sci. **142**(3) (2025)
43. Salimian, S., Mousavi, S.M., Antucheviciene, J.: An interval-valued intuitionistic fuzzy model based on extended VIKOR and MARCOS for sustainable supplier selection in organ transplantation networks for healthcare devices. Sustainability **14**(7), 3795 (2022)
44. Ghosh, D., Dey, S., Pal, A.: A novel multi-criteria decision making approach using trapezoidal intuitionistic fuzzy number. In: 2018 Fourth International Conference on Research in Computational Intelligence and Communication Networks (ICRCICN), pp. 231–236 (2018).https://doi.org/10.1109/ICRCICN.2018.8718706
45. Zhu, L., Meng, X., Shi, Z., Huang, Y.: Selection of prefabricated concrete factories' location based on triangular fuzzy numbers and fuzzy group decision-making. In: 2018 3rd International Conference on Computational Intelligence and Applications (ICCIA), pp. 26–30 (2018). https://doi.org/10.1109/ICCIA.2018.00013
46. Qahtan, S., Sharif, K.Y., Zaidan, A.A., Alsattar, H.A., Albahri, O.S., Zaidan, B.B., Zulzalil, H., Osman, M.H., Alamoodi, A.H., Mohammed, R.T.: Novel multi security and privacy benchmarking framework for blockchain-based IoT healthcare Industry 4.0 systems. IEEE Trans. Ind. Inf. **18**(9), 6415–6423 (2022). https://doi.org/10.1109/TII.2022.3143619
47. Ayağ, Z.: A comparison study of fuzzy-based multiple-criteria decision-making methods to evaluating green concept alternatives in a new product development environment. Int. J. Intell. Comput. Cyber. **14**(3), 412–438 (2021)
48. Chen, Y.-S., Chuang, H.-M., Sangaiah, A.K., Lin, C.-K., Huang, W.-B.: A study for project risk management using an advanced MCDM-based DEMATEL-ANP approach. J. Ambient. Intell. Humaniz. Comput. **10**, 2669–2681 (2019)
49. Thanh, N.V.: Designing a MCDM model for selection of an optimal ERP software in organization. Systems **10**(4), 95 (2022)
50. Karbassi Yazdi, A., Muneeb, F.M., Wanke, P.F., Figueiredo, O., Mushtaq, I.: Critical success factors for competitive advantage in Iranian pharmaceutical companies: a comprehensive MCDM approach. Math. Probl. Eng. **2021**(1), 8846808 (2021)
51. Schaefer, J.L., Siluk, J.C.M., de Carvalho, P.S.: An MCDM-based approach to evaluate the performance objectives for strategic management and development of energy cloud. J. Clean. Prod. **320**, 128853 (2021)
52. Xiao, L., Huang, G., Zhang, G.: Toward an action-granularity-oriented modularization strategy for complex mechanical products using a hybrid GGA-CGA method. Neural Comput. Appl. **34**(8), 6453–6487 (2022)

53. Abbas, S., Alnoor, A., Yin, T.S., Sadaa, A.M., Muhsen, Y.R., Khaw, K.W., Ganesan, Y.: Antecedents of trustworthiness of social commerce platforms: a case of rural communities using multi group SEM & MCDM methods. Electron Commer Res Appl 101322 (2023)
54. Alnoor, A., Abbas, S., Khaw, K.W., Muhsen, Y.R., Chew, X.: Unveiling the optimal configuration of impulsive buying behavior using fuzzy set qualitative comparative analysis and multi-criteria decision approach. J. Retail. Consum. Serv. **81**, 104057 (2024)
55. Cheung, G.W., Cooper-Thomas, H.D., Lau, R.S., Wang, L.C.: Reporting reliability, convergent and discriminant validity with structural equation modeling: a review and best-practice recommendations. Asia Pac. J. Manage. **41**(2), 745–783 (2024)
56. Hair Jr, J.F., Hult, G.T.M., Ringle, C.M., Sarstedt, M., Danks, N.P., Ray, S., Hair, J.F., Hult, G.T.M., Ringle, C.M., Sarstedt, M.: An introduction to structural equation modeling. In: Partial Least Squares Structural Equation Modeling (PLS-SEM) Using R: A Workbook, pp. 1–29. (2021)
57. Kang, H., Ahn, J.-W.: Model setting and interpretation of results in research using structural equation modeling: a checklist with guiding questions for reporting. Asian Nurs. Res. **15**(3), 157–162 (2021)
58. Alsalem, M.A., Alamoodi, A.H., Albahri, O.S., Dawood, K.A., Mohammed, R.T., Alnoor, A., Zaidan, A.A., Albahri, A.S., Zaidan, B.B., Jumaah, F.M., Al-Obaidi, J.R.: Multi-criteria decision-making for coronavirus disease 2019 applications: a theoretical analysis review. Artif. Intell. Rev. (2022). https://doi.org/10.1007/s10462-021-10124-x
59. Salih, M.M., Zaidan, B.B., Zaidan, A.A.: Fuzzy decision by opinion score method. Appl. Soft Comput. **96**, 106595 (2020). https://doi.org/10.1016/j.asoc.2020.106595
60. Khaw, K.W., Camilleri, M., Tiberius, V., Alnoor, A., Zaidan, A.S.: Benchmarking electric power companies' sustainability and circular economy behaviors: using a hybrid PLS-SEM and MCDM approach. Environ. Dev. Sustain. **26**(3), 6561–6599 (2024)
61. Sadaa, A.M., Ganesan, Y., Khaw, K.W., Alnoor, A., Abbas, S., Chew, X., Bayram, G.E.: Based on the perception of ethics in social commerce platforms: adopting SEM and MCDM approaches for benchmarking customers in rural communities. Curr. Psychol. **42**(35), 31151–31185 (2023)
62. Albahri, A.S., Alnoor, A., Zaidan, A.A., Albahri, O.S., Hameed, H., Zaidan, B.B., Peh, S.S., Zain, A.B., Siraj, S.B., Alamoodi, A.H.: Based on the multi-assessment model: towards a new context of combining the artificial neural network and structural equation modelling: a review. Chaos Solitons Fractals **153**, 111445 (2021)
63. Edward, J.L., Kaliyaperumal, P.: Optimizing E-commerce selection under uncertainty: a new framework for decision-making with extended fuzzy TOPSIS. Int. J. Math. Math. Sci. **2024**(1), 4284997 (2024)
64. Sun, H., Zhang, B., Ni, W.: A hybrid model based on SEM and fuzzy TOPSIS for supplier selection. Mathematics **10**(19), 3505 (2022)
65. Chen, H.-M., Wu, H.-Y., Chen, P.-S.: Innovative service model of information services based on the sustainability balanced scorecard: applied integration of the fuzzy Delphi method, Kano model, and TRIZ. Expert Syst. Appl. **205**, 117601 (2022)
66. Roldan Lopez de Hierro, A.F., Sánchez, M., Puente-Fernández, D., Montoya-Juárez, R., Roldán, C.: A fuzzy delphi consensus methodology based on a fuzzy ranking. Mathematics **9**(18), 2323 (2021)
67. Ahmad Basri, M.A.F., Wan Ismail, W.S., Kamal Nor, N., Mohd Tohit, N., Ahmad, M.N., Mohamad Aun, N.S., Mohd Daud, T.I.: Validation of key components in designing a social skills training content using virtual reality for high functioning autism youth—a Fuzzy Delphi method. PLoS ONE **19**(4), e0301517 (2024)
68. Nuriyev, A.M.: Fuzzy MCDM models for selection of the tourism development site: the case of Azerbaijan. F1000Res. **11**, 310 (2022). https://doi.org/10.12688/f1000research.109709.1
69. Moradpour, N., Pourahmad, A., Ziari, K., Hataminejad, H., Sharifi, A.: Downscaling urban resilience assessment: a spatiotemporal analysis of urban blocks using the fuzzy Delphi method and K-means clustering. Build. Environ. **263**, 111898 (2024)

Chapter 11
Machine Learning in Business Research

Abstract This chapter outlines the use of machine learning and metaheuristic algorithms to help address complex problems in business. When information grows and changes quickly, traditional tools are usually unable to deal with all the issues businesses face. Machine learning gives effective abilities in recognizing patterns, predicting outcomes, and automating tasks, while metaheuristic algorithms offer useful and flexible ways to solve problems without a straightforward answer. In the end, this chapter gives readers a detailed picture of how intelligent systems can assist businesses in making decisions.

Learning Objectives

1. Identify the significance of Metaheuristic Algorithms.
2. Introduce how metaheuristics are used to improve machine learning parameters or structures.
3. Explore new developments in combining machine learning, optimization, and business intelligence.

11.1 Machine Learning

The rapid advancements in information, data technology, and data analysis have greatly increased the use of Machine learning (ML) techniques across many fields [1, 2]. Business and financial applications using ML such as marketing, e-commerce and energy, have advanced a lot and the outcomes have been very promising [3]. ML is everywhere in the business world today, supporting software in marketing, finance, supply chains, and the energy field. Recent work indicates that business analytics, stock prediction, demand forecasting, and energy management are improved by using ML techniques, especially deep neural networks [4]). ML's flexibility makes it possible for businesses to study huge amounts of both structured and unstructured data, leading to good results in e-commerce, risk management, and managing

X. Chew et al., *Partial Least Squares Structural Equation Modeling and Complementary Methods in Business Research*, Information Systems Engineering and Management 67, https://doi.org/10.1007/978-3-032-01055-1_11

customer groups [5]. It can capture difficult patterns and improving forecasting results, ML models in finance and operations, which cover convolutional and recurrent networks, are now widely used [6]. ML approaches outperform most econometric techniques in the use of large various types of data and in quick decision-making and forecasting. These benefits come from the fact that ML models can train without making fixed assumptions and ensure accurate predictions of the key results [7]. Even though ML models have achieved many successes, practitioners working with them have noticed important obstacles when used in business situations [2].

Certain inherent problems with ML can lead to poor results and limited generalization in practical business use [8]. Many models become overfitted, underfitted, or have poor generalization, which leads to lower accuracy when applied to fresh big data. Generalization, in particular, is considered a significant issue in ML for business and finance. A major problem faced in deep learning is local optima. Standard gradient-based training easily gets trapped in suboptimal solutions and needs a cost function that is both smooth and can vary slowly [9]. However, handling the multimodal or discontinuous real-life objectives we often see may be difficult for this type of training. A study finds that gradient techniques for neural networks commonly get caught in narrow solutions and consume a lot of resources [10]. Also, the troubles of setting up the learning rate, regularization weight, and architecture size are very similar and involve solving a difficult, non-convex problem with lots of different settings [11]. Selecting the optimal set of hyperparameters is commonly needed in machine learning because it can have a strong impact on a model's results [12, 13]. It often becomes a challenge to interpret the results (users need to understand the models) and to manage data problems (lack of good data limits how much can be learned). In short, ML models often require a significant amount of data and are finicky about how they are configured, problems that often appear are overfitting, getting stuck in a local minimum, and the difficulty of choosing multiple hyperparameters [2, 11, 13].

11.2 Metaheuristic Optimization Algorithms

Over recent years, it has become necessary to use more effective optimisation techniques, including meta-heuristic algorithms, because the problems faced in practice have become more elaborate and harder to solve. Most of these algorithms generate stochastic values that are close to the optimum [11, 14]. It is often difficult to find solutions to realistic problems because of their complex, nonlinear nature and large range of possibilities [15].

Both exploration and exploitation searches are important types used by metaheuristics [16]. Exploration is the skill needed to find new areas everywhere in the search space. This ability helps us to escape from stagnation at a local minimum. Exploitation, then, means trying to find local answers with the goal of enhancing them [17]. A strong balance between the two approaches is crucial for the success of the algorithm [18]. Multiple meta-heuristic types can be found such as those inspired by evolutionary algorithms, swarm intelligence, simulation of physical systems and human methods [11].

The topic of theoretical research in the literature is sorted into three categories: increasing the performance of existing algorithms, putting together several algorithms, and adding new ones. There is a great deal of research happening in these parts, and an abundance of algorithms and software is being made. Researchers are unable to use only one algorithm because the No Free Lunch theorem proves that no single optimisation algorithm is suited for every issue [19]. Based on our research, Ridha et al. introduced GCAOAEmNR [20], a modified third-order Newton–Raphson algorithm backed by a convergence result, to use in energy prediction models to benefit exploration and exploitation. As well, researchers in 2022 [21] introduced the thought that mixing many highly effective hybrid techniques is key to getting good results from the GCAOAAdLM model. A specific goal function is developed to support the optimal selection of initial root parameters for the double diode and three diode models, using Levenberg–Marquardt with an adaptive damping parameter as inspiration. The experiments demonstrated that our approach helped to significantly reduce the RMSE and the MBE. There are four main groups of meta-heuristic algorithms: those based on swarm intelligence, evolutionary computation, physics and hybrid methods [11]. Section 2.1 will explain some of swarm algorithms, Sect. 2.2 about evolutionary, Sect. 2.3 explains physics, and Sect. 2.4 hybrid meta-heuristic.

11.2.1 Swarm Intelligence (SI)

Swarm intelligence describes a system in which many simple agents work together locally and arrange themselves to solve big problems [22]. The idea is partially based on groups in nature such as ants, bees, birds, and fish, where everyone in the group acts intelligently and flexibly, even though one person doesn't control all activities. Agents in the SI system have straightforward rules and mostly share information without direct contact and as a group, they solve complicated problems effortlessly [23]. Solving hard computational problems has been possible with swarm intelligence, as it is used in robotics, data mining, telecommunications, and optimization areas [24]. Under this category, you will see many algorithms, with some of them discussed further in the section below.

11.2.1.1 Particle Swarm Optimization (PSO)

PSO uses behavior found in bird flocking and fish schooling. The population consists of many possible solutions [25]. In each move, a particle changes its place in the search space using its own personal best result and the swarm's global best position. Intuitively, since each particle updates its velocity with reference to the best solutions, the swarm is guided to areas with favourable solutions. PSO is easy to put into practice and enjoys widespread popularity due to it does well in complex or multimodal scenarios [26]. Many researchers used PSO since it includes only a few needed parameters.

11.2.1.2 Shuffled Frog-Leaping Algorithm (SFLA)

The Shuffled Frog-Leaping Algorithm (SFLA) uses both the principles of memetic evolution and the way swarm life forms cooperate. In 2003, Eusuff and Lansey [27] introduced the algorithm which takes its name from frogs in a swamp, since groups of frogs exchange messages and take action to find the richest part of the swamp [27]. SFLA combines the worldwide search of genetic algorithms with the local search of PSO, letting it solve both types of optimization problems efficiently.

In the SFLA, the group of frogs, representing potential solutions, is put in the multidimensional space of the search. Frogs start out by accident, laying out various options for solutions. All individuals are grouped into subsets called memeplexes, and each search for the best solutions by itself. In each memeplex, frogs are influenced by each other through the guidance of the best frog (the leader), which tries to raise the position of the worst one. Memeplexes are mixed together from time to time, which ensures frogs share information and challenge themselves by visiting different areas of the search space [28].

11.2.1.3 Ant Colony Optimisation (ACO)

An idea for optimising process parameters was introduced by Ali Dorigo et al. [29]. Foraging methods of several ant species have inspired the creation of ACO. To help the colony's ants, these ants leave pheromone marks in the soil, pointing out a good way to travel. A similar technique is applied in ant colony optimization for solving complex optimisation issues. ACO operates in discrete as well as continuous spaces to solve different static as well as dynamic optimization concerns with the help of a colony of artificial ants. The system divides the work among artificial ants that are very simple. They cooperate covertly by leaving pheromone trails. Agent-based models that use this approach alternate between iterations based on a probability distribution [30].

11.2.1.4 Grasshopper Optimisation Algorithm (GOA)

Saremi and his colleagues were the first to introduce GOA [31]. By imitating how grasshoppers move, it is used to solve several everyday optimisation challenges. All grasshoppers form a network, and this network links them so they can move in unison and respond to changes. Network members can work together to agree on who will be prey for the group. Grasshoppers are affected by gravity and by repulsive forces acting on them. Grasshoppers can look for and find suitable homes because the repelling force helps, and they also go to different areas because of the gravitational force.

When gravity is the same as the repulsive force, the place where a grasshopper sit is called the optimal adaptation region. Because they were unaware of the exact place to aim, the best area for adaptation was selected as the target position. The grasshopper will go towards where the network wants it to go. Grasshoppers' exploration regions are kept adjusted by the system to ensure equal attention to the entire population and specific swarms. Gradually, all the grasshoppers will come together and work out the best option.

11.2.1.5 Firefly Algorithm (FA)

The firefly's elegant behaviour was the reason FA Yang was designed [32]. How fireflies light up forms the basic idea behind the FA. According to the FA method, the solution space is filled with fireflies, each one of which represents a possible starting point. Fitness score for a firefly is given based on its target function, and then a light intensity is assigned to it. Lighter next to a firefly will draw in other fireflies that are less bright. As the optimization needs evolve, algorithms like FA will still be important for artificial intelligence, engineering, and business matters [33].

11.2.1.6 A Technique Known as the Salp Swarm Algorithm (SSA)

The Foraging and Migration of Salps inspired Mirjalili and his colleagues to introduce the SSA [34]. These marine Salpidae are tunicates called salps, and their jellylike bodies, which are transparent, give them the look of jellyfish. Moving through water, they do so by swallowing it, expelling, and jetting backwards, which gives them the power to drift or swim. It is common for salps to join into groups that move and eat more easily because there are many together. Within salp chains, the first salp explores the environment ahead, then others adjust their places only after seeing the direction of the leading salp.

In SSA, these behavior patterns are modeled using math to address continuous nonlinear problems. The group of candidate solutions is separated into leaders (the best ones currently) and followers. It finds its new position in relation to the global solution (the food source) by using both exploration and exploitation metrics, which are generated at random. Salp followers apply a simplified Newtonian model that uses the leader's immediate position to figure out how to move.

11.2.1.7 Grey Wolf Optimiser Algorithm (GWO)

The GWO was first designed by Mirjalili et al. [35] and is inspired by the social order and eating habits of grey wolves (Canis lupus). The algorithm sorts the solutions in the population based on the pack's leadership, so the solutions are grouped as alpha, beta, delta, and omega, which are the top, second, third, and remaining solutions.

Hierarchical organization guides the way individuals deal with the search for solutions within the problem space. The main aim of the GWO algorithm is to imitate the behaviors of grey wolves while hunting together, such as circling the prey, attacking, and hunting for it. Finding the best three values among the wolves (α, β, and δ) helps the search. The others adjust their position, which helps the algorithm get close to the top results.

GWO is especially well suited because it detects global optimums by using a balanced mixture of broad search and targeted search, which is updated at each new iteration. It works well on optimization problems that involve nonlinear, multiple modes, and high dimensions, since it provides better estimates and converges faster than a number of classical optimization methods. Therefore, GWO remains one of the most popular bio-inspired algorithms, especially these days in engineering design, machine learning, image processing, power systems, and business optimization [36]. Due to being simple, performing smoothly on computers, and success in escaping from local best solutions, the GWO algorithm remains popular in handling real-world optimization cases.

While the GWO, FA, and SSA are commonly used, other swarm-based approaches can also help in business applications. The Artificial Bee Colony (ABC), Bat Algorithm, Cuckoo Search, and Whale Optimization Algorithm (WOA) all belong to this group. The popularity of each algorithm depends on the strengths and weaknesses it has, which are important for selecting the best one for a particular business problem.

11.2.2 Evolutionary Metaheuristic Algorithms

Evolutionary metaheuristic algorithms are methods of solving difficult optimization problems that copy elements of natural evolution. When traditional optimization fails, such as in high-dimensional, nonlinear, discontinuous, or multi-modal spaces, these algorithms are very useful. Based on Darwin's concepts of natural selection, mutation, crossover, and survival of the fittest, evolutionary algorithms keep modifying a set of possible answers so that they move closer to the best solution [37].

Some of the best-known examples of evolution-based algorithms are the Genetic Algorithm, Differential Evolution, Invasive Weed Optimisation, and Cultural Algorithms. These algorithms try out whole groups of "people" as possible solutions using a technique called population-based search. Applying selection, crossover, and mutation to the population on each generation helps it find better solutions [38, 39]. Next sections will explain some of popular algorithms.

11.2.2.1 Genetic Algorithm (GA)

The GA of Holland [40] is shown to be reliable, strong, and well-adjusted by following natural selection and evolution. The principles behind natural biological evolution formed the GA, and it is widely used to solve optimisation problems

[41]. In essence, the method starts by creating a set of genes randomly to serve as possible answers to a problem. For each chromosome, the fitness function that shows the chances of being selected must be set up. In crossover, distinctive parts of two different chromosomes are swapped to form a better progeny. Therefore, genes at random points along several chromosomes are changed. The final genetic modification scientists perform is called a "mutation" [42].

11.2.2.2 Differential Evolution (DE)

Storn and Price introduced DE, a flexible algorithm that makes it easy to solve complex continuous optimization problems [43]. DE works with a group of candidate answers and uses vector operations to explore all parts of the solution space. DE is distinct from traditional evolutionary algorithms because it mainly uses differential variation to find new solutions.

DE tries to create new solutions (trial vectors) by merging parts of existing individuals in the group. It is normally performed by selecting two random individuals and subtracting their weighted difference from a third individual. A new individual is created and is compared to the original (parent) one. The new individual takes the place of the original only when it is fitter. This method is called a greedy selection strategy which makes certain that the population either grows better or stays the same in each iteration. According to [44], due to DE uses a greedy strategy, it usually outperforms other optimization methods in terms of how quickly it converges and how accurate its answers are.

The operation of DE in particle swarm optimization depends on three control factors: the mutation rate (F), the crossover rate (CR), and the size of the population (NP). DE is very simple, but still shows strong abilities to find the best solutions worldwide and perform well in application areas such as engineering, learning, machine training, portfolio growth, and supply chain management. Like many metaheuristics, DE encounters difficulties such as quick convergence or stopping in a limited range in some high-dimensional or misleading regions. Many different versions and mixed approaches of DE have been designed to make the method more flexible and sturdier [45].

11.2.2.3 Invasive Weed Optimization (IWO)

Mehrabian and Lucas first introduced the IWO algorithm as an evolutionary algorithm based on nature [46]. The open-source approach gets its name from the way invasive weeds in nature grow very fast, reproduce quickly, and can thrive in most places. Ravenfeld used weeds as a symbol since they fit a wide range of environments and usually outcompete other plants, so they make a good example for designing effective and adaptable optimization algorithms.

A set of randomly generated individuals called seeds is used as the first phase in IWO. They are regarded as solutions, allowed to grow and reproduce as much as they can, like weeds in nature. With time, those who do well in life make more offspring, whereas those who do poorly may have fewer offspring or disappear from the population, much like natural selection [47]. One feature that defines IWO is using spatial dispersal, in which new seeds are positioned around the earlier parents by a random function in a bell-shaped curve. As the algorithm keeps running, the spread in the distribution becomes smaller, which makes it easier for the algorithm to switch from exploring widely to searching more specifically [48].

Basic but important rules for weed populations in nature, such as how seeds are formed, how they are spread, the battles over resources, and the fittest, are all used to form the algorithm's rules. They allow IWO to work effectively, quickly, and reliably in finding the best or nearly best solutions among many possible paths in a search area [49].

11.2.2.4 Biogeography-Based Optimization (BBO)

BBO is an algorithm that takes its idea from how species in nature are spread across different locations. Dan Simon [50] first suggested the algorithm, using the way real species migrate, adapt, and change across secluded spots as a source of inspiration. BBO is mainly founded on the idea of island biogeography, which is concerned with the balance between immigration and extinction of species in habitats.

Every hypothesis in BBO becomes a habitat, and how suitable a habitat is for the problem depends on its Habitat Suitability Index (HSI), related to the fitness value from standard optimization. Such good habitats are good solutions and tend to share their characteristics with others, much like species that move from suitable areas to less suitable ones. Low-HSI areas are usually able to accept new developments, which improves their overall quality [51].

The two major operations in evolution in BBO are migration and mutation. It models how features from the best habitats are transferred to the less suitable areas in order to create new offspring. Mutation adds some randomness so diversity is maintained, and early convergence does not occur. Because of this process, the population grows to find better solutions, but ensures there is a proper mix between searching for new answers and making use of ones that have been found [52]. BBO is successfully employed in engineering, power, recognition, machine learning, and decision-making contexts.

11.2.3 Physics-Based Algorithms (PH)

PH belong to a group of optimization approaches developed based on principles and laws from physics regarding the movement of matter and energy. Unlike bio-inspired methods inspired by organisms, physics-based algorithms derive inspiration from ideas like gravity, electromagnetism, thermodynamics, motion and field interactions.

They take physical ideas and turn them into mathematical formulas to help search through complicated solutions [53].

Using force-based interactions, like in physics, happens frequently in PH algorithms by guiding agents or particles. A major benefit of physics-based algorithms is that they naturally keep a good balance between searching for new information and choosing the best actions, guided by physical rules. Thus, they usually handle complex, multi-peaked and high-dimensional problems better, without getting trapped in local optima. Their straightforward design, many adjustments features and ability to fit many applications are what make them valuable [54]. Some of popular PH algorithms will explain below.

11.2.3.1 Multi-verse Optimizer (MVO)

The MVO is an algorithm based on science that was first developed by Mirjalili et al. [55] after they drew inspiration from white holes, black holes and wormholes. The ideas are based on theories in astrophysics and are represented mathematically in MVO so that exploration, exploitation and local search can take place in the solution space.

MVO refers to universes instead of solutions and the metaphorical way to measure their quality (fitness) is by measuring their inflation rate. Using white holes, solutions of high quality can share their qualities which helps the search process by spreading effective features across all solutions. Taking the reverse approach, black holes take in bad solutions and push unfit candidates to the side, reducing their influence on the result. Also, wormholes are considered because they allow candidate solutions to get nearer to the known best solution to improve it [56].

11.2.3.2 Water Cycle Optimization Algorithm (WCA)

Eskandar et al. [57] invented the WCA, a metaheuristic that takes inspiration from the water cycle in the environment. It uses an algorithm to show how water moves downstream in rivers and streams and evaporates, precipitates and flows out as runoff in nature. Water Ca comes up with solutions by mimicking the movement of water particles downhill until they reach the sea which is like reaching the optimum answer.

In the WCA, every candidate solution is seen as a water droplet and all together they become the sea, with the best solution known as its surface. The possible solutions that come next are considered to be rivers and the final ones, the ones that fall short, are called streams. Solutions are also found in streams, then in rivers and finally in the sea which illustrates a gradual improvement. When ideas are being tested, the most effective solutions gather others, much as water goes to the lowest spot. When a group of solutions is stagnating, the algorithm adds new possible solutions by using evaporation and rainfall mechanisms [58].

11.2.4 Hybrid Meta-Heuristic Algorithms

Hybrid algorithms include combining two or more optimization methods which work jointly to boost performance, enhance accuracy and overcome difficulties faced by sole techniques. The goal is for hybrid systems to use the strengths of each algorithm and make up for their flaws, helping solve problems better. Usually, the aim of designing hybrid algorithms is to make convergence quicker, the solution more precise and the approach stronger in handling a wide array of optimization tasks [59].

The premise of hybridization is that no algorithm is best for every type of problem, as the No Free Lunch (NFL) theorem tells us. Therefore, hybrid methods try to blend the ability of one algorithm (such as a global search) with another algorithm's ability to refine results locally, leading to a stronger and more adaptable optimization system. By using both a population-based algorithm (PSO) and a local search algorithm (SA), problems like stopping early and getting stuck at local optima can be avoided.

A number of studies such as [60, 61] have confirmed that hybrid algorithms are more efficient and accurate than others. It is often hard for traditional methods to solve high-dimensional, multi-modal and dynamic problems, but these methods can help. As [59] states, using hybrid techniques who addresses randomization inefficiency, less than enough intensity and issues that involve being trapped in local minima makes them suitable for real-life scenarios.

11.3 Metaheuristic with Machine Learning Algorithms

Adopting metaheuristics can address the shortcomings that can affect ML models. Hyperparameter optimization (HPO) is a major area to focus on. Metaheuristics can search the space of hyperparameters in a more effective way. As an illustration, Vincent and [13] found that combining Bayesian optimization with evolutionary strategies (using Genetic Algorithms, Differential Evolution, and CMA-ES) helped in tuning image classification models. It is commonly used to say that PSO and ACO are great for finding the best hyperparameter values for different problems, especially when the search space is broad and exploring it can be costly. On the other hand, simple grid or random search might not explore all the important areas, but population-based metaheuristics keep using methods that allow efficient exploration and detailed study of the hyperparameter landscape [10]. Model training and architecture design may benefit from using metaheuristics. Hybrid methods in neural networks swap out or combine gradient descent with updates that are driven by evolution. An example is training network weights or even restructuring networks with GA or particle-swarm, which helps prevent being caught in local minima. As per the review, these algorithms accurately adjust the parameters of deep models (weights, layers, and learning rates) and using hybrid methods increases the algorithms' performance and enables them to sort out difficult optimization problems.

So, an example of this is that a GA can change the network's parameters alongside backpropagation, making training speed up and improving the network's level of generalization. Also, swarm algorithms such as Ant Colony have been applied to picking a suitable group of input features that reduces the risk of overfitting and makes the model simpler (a well-known approach used in pattern classification tasks). Results from practical research regularly show that using metaheuristics with ML models is more effective [62]. Often, PSO or ACO performs better than manually or exhaustively adjusting the hyperparameters of a support vector machine or a neural network. According to one scientific document, these metaheuristic approaches are important when the data space is difficult or when evaluating the objective function is challenging or inconvenient, which is how most business situations happen to be. Since they expand the search and help prevent overfitting, these techniques create models that perform better on unknown data. In addition, metaheuristics can help automate the challenging process of parameter tuning, which makes ML simpler for people to apply. Overall, metaheuristic optimization provides a useful set of tools for handling typical problems in ML. Integrating GA, PSO, and ACO algorithms allows businesses to strengthen their hybrid systems, neural networks, and other learners by changing their architecture and parameters, which improves their performance [63]. Such mixed techniques speed up training, get out of poor minima, and make models that do better generalization than those trained only using gradients. As a consequence, metaheuristics are now frequently used in applied ML, playing a role in fully utilizing the benefits of machine learning in the areas of business and finance.

11.4 Future Direction

As businesses encounter more and more complex and information-rich conditions, using ML and metaheuristic optimization techniques can help them make better decisions, anticipate future results, and achieve better operational performance. There are many promising upcoming areas where these technologies are predicted to have a huge impact.

11.4.1 Automated Decision Intelligence is Used by Businesses

We can expect automated decision intelligence to play a major role in business activities in the following years. These systems will apply machine learning for making predictions and classifying data, and they will rely on metaheuristic algorithms to improve decisions on topics such as controlling inventory, handling logistics, and setting dynamic prices under set restrictions. Using these technologies together will allow for real-time, flexible decision-making engines that get smarter with each bit of data.

11.4.2 Hybrid and Adaptive Types

More hybrid systems that mix machine learning models (neural networks, decision trees, or ensembles) with metaheuristic algorithms (PSO, GA, GWO, DE) are likely to appear. They can automatically adjust attributes of ML models such as parameters, features, or the way the model is structured, which improves how it works in general. Future studies will concentrate on how hybrid systems adjust their methods on their own depending on the complexity of a problem and the data involved.

11.4.3 AI that Explains and Upholds Ethical Principles in Optimization

Because AI affects business decisions more now, people are requiring explainable and ethical algorithms. Using interpretability in ML and metaheuristic-based systems will play a key role in achieving transparency, trust, and abiding by rules. For this reason, new optimization models and systems that let decision-makers be closely involved in the optimization process will be developed.

11.4.4 Big Data Can Use Scalable and Distributed Metaheuristic Algorithms

Because of the use of big data in business analytics, scalable metaheuristic algorithms that work with parallel, distributed, or cloud infrastructure will become more important. Businesses will now be able to make optimized decisions on large sets of data in practically real-time. This type of research will explore decentralized and federated metaheuristics combined with edge computing and privacy protection.

By linking artificial intelligence, operational research, and business strategy, these new directions open up many attractive opportunities for both academics and professionals. Due to rapid improvements in algorithms, many fields will experience more innovation, stability, and competitiveness worldwide.

References

1. Al-hchaimi, A.A.J., Alomari, M.F., Muhsen, Y.R., Sulaiman, N.B., Ali, S.H.: Explainable machine learning for real-time payment fraud detection: building trustworthy models to protect financial transactions. In: International Conference on Explainable Artificial Intelligence in the Digital Sustainability, pp. 1–25 (2024)

2. Liaras, E., Nerantzidis, M., Alexandridis, A.: Machine learning in accounting and finance research: a literature review. Rev. Quant. Financ. Acc.Financ. Acc. (2024). https://doi.org/10.1007/s11156-024-01306-z
3. Athey, S., Imbens, G.W.: Machine learning methods that economists should know about. Annu. Rev. Econ. **11**(1), 685–725 (2019)
4. Chen, H., Fang, X., Fang, H.: Multi-task prediction method of business process based on BERT and transfer learning. Knowl.-Based Syst..-Based Syst. **254**, 109603 (2022)
5. Kareegalan, K., Asmawi, A., Abdullah, M.T., Ninggal, M.I.H., Abdullah, M.D.H., Muhsen, Y.R.: Convolutional Long Short-Term Memory for Fileless Malware Detection (n.d.)
6. Chen, Z., Chen, W., Shi, Y.: Ensemble learning with label proportions for bankruptcy prediction. Expert Syst. Appl. **146**, 113155 (2020)
7. Nazareth, N., Reddy, Y.V.R.: Financial applications of machine learning: a literature review. Expert Syst. Appl. **219**, 119640 (2023)
8. Rajvanshi, S., Kaur, G., Dhatwalia, A., Arunima, S.A., Bhasin, A.: Research on problems and solutions of overfitting in machine learning. In: International Conference on Artificial-Business Analytics, Quantum and Machine Learning, pp. 637–651 (2023)
9. Roelofs, R., Shankar, V., Recht, B., Fridovich-Keil, S., Hardt, M., Miller, J., Schmidt, L.: A meta-analysis of overfitting in machine learning. Adv. Neural Inf. Process. Syst. **32** (2019)
10. Kaveh, M., Mesgari, M.S.: Application of meta-heuristic algorithms for training neural networks and deep learning architectures: a comprehensive review. Neural. Process. Lett. **55**(4), 4519–4622 (2023)
11. Khairan, H.E., Zubaidi, S.L., Muhsen, Y.R., Al-Ansari, N.: Parameter optimisation-based hybrid reference evapotranspiration prediction models: a systematic review of current implementations and future research directions. Atmosphere **14**(1), 77 (2022)
12. Muhsen, Y.R., Husin, N.A., Zolkepli, M.B., Manshor, N., Al-Hchaimi, A.A.J., Ridha, H.M.: Enhancing NoC-based MPSoC performance: a predictive approach with ANN and guaranteed convergence arithmetic optimization algorithm. IEEE Access (2023)
13. Vincent, A.M., Jidesh, P.: An improved hyperparameter optimization framework for AutoML systems using evolutionary algorithms. Sci. Rep. **13**(1), 4737 (2023)
14. Methods, C., Mech, A., Abualigah, L., Diabat, A., Mirjalili, S., Abd, M.: ScienceDirect the arithmetic optimization algorithm. Comput. Methods Appl. Mech. Eng.. Methods Appl. Mech. Eng. **376**, 113609 (2021). https://doi.org/10.1016/j.cma.2020.113609
15. Abualigah, L.: Multi-verse optimizer algorithm: a comprehensive survey of its results, variants, and applications. Neural Comput. Appl.Comput. Appl. **32**(16), 12381–12401 (2020)
16. Abualigah, L., Diabat, A., Geem, Z.W.: A comprehensive survey of the harmony search algorithm in clustering applications. Appl. Sci. **10**(11), 3827 (2020)
17. Abualigah, L.: Group search optimizer: a nature-inspired meta-heuristic optimization algorithm with its results, variants, and applications. Neural Comput. Appl.Comput. Appl. **33**(7), 2949–2972 (2021)
18. Mohammed, H., Hizam, H., Mirjalili, S., Lutfi, M., Effendy, M., Ahmadipour, M.: Parameter extraction of single, double, and three diodes photovoltaic model based on guaranteed convergence arithmetic optimization algorithm and modified third order Newton Raphson methods. Renew. Sustain. Energy Rev. **162**(March), 112436 (2022). https://doi.org/10.1016/j.rser.2022.112436
19. Wolpert, D.H., Macready, W.G.: No free lunch theorems for optimization. IEEE Trans. Evol. Comput.Evol. Comput. **1**(1), 67–82 (1997)
20. Ridha, H. M., Hizam, H., Mirjalili, S., Othman, M. L., Ya'acob, M. E., & Ahmadipour, M. (2022). Parameter extraction of single, double, and three diodes photovoltaic model based on guaranteed convergence arithmetic optimization algorithm and modified third order Newton Raphson methods. Renew. Sustain. Energy Rev. **162**(March), 112436. https://doi.org/10.1016/j.rser.2022.112436
21. Ridha, H.M., Hizam, H., Mirjalili, S., Othman, M.L., Ya'acob, M.E., Ahmadipour, M., Ismaeel, N.Q.: On the problem formulation for parameter extraction of the photovoltaic model: novel integration of hybrid evolutionary algorithm and Levenberg Marquardt based on adaptive

damping parameter formula. Energy Convers. Manage. **256**(February), 115403. https://doi.org/10.1016/j.enconman.2022.115403
22. Metcalf, L., Askay, D.A., Rosenberg, L.B.: Keeping humans in the loop: pooling knowledge through artificial swarm intelligence to improve business decision making. Calif. Manage. Rev. **61**(4), 84–109 (2019)
23. Kelly, R.: The Nature of Business Transformation: A Swarm Intelligent Approach to Reinventing Organisations. Routledge (2022)
24. Tang, J., Liu, G., Pan, Q.: A review on representative swarm intelligence algorithms for solving optimization problems: applications and trends. IEEE/CAA J. Autom. Sinica **8**(10), 1627–1643 (2021)
25. Gao, X., Liang, L.: Enterprise business management resource adaptive scheduling algorithm based on particle swarm optimization. Int. J. High Speed Electron. Syst. 2540538 (2025)
26. Chen, M., Du, W.: Dynamic relationship network and international management of enterprise supply chain by particle swarm optimization algorithm under deep learning. Expert. Syst. **41**(5), e13081 (2024)
27. Eusuff, M., Lansey, K., Pasha, F.: Shuffled frog-leaping algorithm: a memetic meta-heuristic for discrete optimization. Eng. Optim.Optim. **38**(2), 129–154 (2006)
28. Sahani, S.K., Chaudhary, A.K., Thapa, G., Nepal, B., Karki, M., Poudel, R.K., Singh, V.V., Sahani, K.: A review of the shuffling frog leading algorithm for production optimization issues in industries is integrated into vertical transportation systems: comprehensive studies of mathematical and statistical problems. J. Inte. Crisis Risk Commun. Res. **7**(2), 552 (2024)
29. Dorigo, M., Maniezzo, V., Colorni, A.: Ant system: optimization by a colony of cooperating agents. IEEE Trans. Syst., Man, Cybern. B **26**(1), 29–41 (1996)
30. Awadallah, M.A., Makhadmeh, S.N., Al-Betar, M.A., Dalbah, L.M., Al-Redhaei, A., Kouka, S., Enshassi, O.S.: Multi-objective ant colony optimization. Arch. Computat. Methods Eng. **32**(2), 995–1037 (2025)
31. Saremi, S., Mirjalili, S., Lewis, A.: Grasshopper optimisation algorithm: theory and application. Adv. Eng. Softw.Softw. **105**, 30–47 (2017)
32. Yang, X.-S.: Firefly algorithm, stochastic test functions and design optimisation. Int. J. Bio-Inspired Comput. **2**(2), 78–84 (2010)
33. Zhang, H.: Study on dynamic allocation method of English teaching resources based on firefly algorithm. Int. J. Bus. Intell. Data Min. **26**(1–2), 218–235 (2025)
34. Mirjalili, S., Gandomi, A.H., Mirjalili, S.Z., Saremi, S., Faris, H., Mirjalili, S.M.: Salp swarm algorithm: a bio-inspired optimizer for engineering design problems. Adv. Eng. Softw.Softw. **114**, 163–191 (2017)
35. Mirjalili, S., Mirjalili, S.M., Lewis, A.: Grey wolf optimizer. Adv. Eng. Softw.Softw. **69**, 46–61 (2014)
36. Makhadmeh, S.N., Al-Betar, M.A., Doush, I.A., Awadallah, M.A., Kassaymeh, S., Mirjalili, S., Zitar, R.A.: Recent advances in Grey Wolf Optimizer, its versions and applications. IEEE Access (2023)
37. Ahmad, M., Javaid, N., Niaz, I.A., Almogren, A., Radwan, A.: A bio-inspired heuristic algorithm for solving optimal power flow problem in hybrid power system. IEEE Access **9**, 159809–159826 (2021)
38. Benaissa, B., Kobayashi, M., Al Ali, M., Khatir, T., Elmeliani, M.E.A.E.: Metaheuristic optimization algorithms: an overview. HCMCOU J. Sci.–Adv. Computat. Struct. 33–61 (2024)
39. Candel, G., Sánchez-Anguix, V., Alberola, J.M., Julián, V., Botti, V.: An evolutionary metaheuristic for forming teams in the classroom with constraints. Neurocomputing **638**, 130138 (2025)
40. Holland, J.H.: Adaptation in Natural and Artificial Systems: An Introductory Analysis with Applications to Biology, Control, and Artificial Intelligence. MIT Press (1992)
41. Lipitakis, A.-D., Lipitakis, E.A.E.C.: Artificial intelligence and business: a hybrid genetic algorithm for e-business strategic planning and performance evaluation. Bus. Manage. Rev. **9**(2), 1–12 (2017)

42. Adnane, Y.I., Zerari, M.: Optimizing business intelligence classification rule mining using quantum-inspired genetic algorithm. IEEE Access (2024)
43. Price, K., Storn, R.M., Lampinen, J.A.: Differential Evolution: A Practical Approach to Global Optimization. Springer Science & Business Media (2006)
44. Eltaeib, T., Mahmood, A.: Differential evolution: a survey and analysis. Appl. Sci. **8**(10), 1945 (2018)
45. Padhy, A.P., Panigrahi, S., Singh, V.P., Pratyasha, P.: Model order reduction for SISO and MIMO system using improved adaptive differential evolution algorithm. Soft Comput. 1–20 (2024)
46. Mehrabian, A.R., Lucas, C.: A novel numerical optimization algorithm inspired from weed colonization. Eco. Inform. **1**(4), 355–366 (2006)
47. Jia, Y.: Chaotic-invasive weed optimization based K-means clustering for identifying financial distress in companies. In: 2025 International Conference on Intelligent Systems and Computational Networks (ICISCN), pp. 1–5 (2025)
48. RJ, K.: Invasive weed optimization with deep transfer learning for multispectral image classification model. Multimedia Tools Appl. **83**(15), 45519–45534 (2024)
49. Koc, I.: A comprehensive analysis of grid-based wind turbine layout using an efficient binary invasive weed optimization algorithm with levy flight. Expert Syst. Appl. **198**, 116835 (2022)
50. Simon, D.: Biogeography-based optimization. IEEE Trans. Evol. Comput.Evol. Comput. **12**(6), 702–713 (2008)
51. Zhang, Y., Gu, X.: A biogeography-based optimization algorithm with local search for large-scale heterogeneous distributed scheduling with multiple process plans. Neurocomputing **595**, 127897 (2024)
52. Garg, V., Deep, K., Kulkarni, A.J.: Laplacian biogeography-based algorithm using a gaining–sharing knowledge-based strategy for global optimization problems and the Lennard-Jones problem. Eng. Optim. 1–32 (2025)
53. Su, H., Zhao, D., Heidari, A.A., Liu, L., Zhang, X., Mafarja, M., Chen, H.: RIME: a physics-based optimization. Neurocomputing **532**, 183–214 (2023)
54. Nguyen, P.K.T., Tran, T.T.H., Nguyen, T.L.: Combination of physics-based model and artificial intelligence for rapid simulation and optimization of dark fermentative hydrogen production from water hyacinth. Int. J. Energy Res. **2024**(1), 5630435 (2024)
55. Mirjalili, S., Mirjalili, S.M., Hatamlou, A.: Multi-verse optimizer: a nature-inspired algorithm for global optimization. Neural Comput. Appl.Comput. Appl. **27**, 495–513 (2016)
56. Jui, J.J., Molla, M.M.I., Ahmad, M.A., Hettiarachchi, I.T.: Recent advances and applications of the multi-verse optimiser algorithm: a survey from 2020 to 2024. Arch. Computat. Methods Eng. 1–34 (2025)
57. Eskandar, H., Sadollah, A., Bahreininejad, A., Hamdi, M.: Water cycle algorithm—a novel metaheuristic optimization method for solving constrained engineering optimization problems. Comput. Struct.. Struct. **110**, 151–166 (2012)
58. Alimorad, H.: Optimization of dynamic control systems using water cycle algorithm. J. Franklin Inst. **361**(8), 106831 (2024)
59. Goodarzian, F., Kumar, V., Abraham, A.: Hybrid meta-heuristic algorithms for a supply chain network considering different carbon emission regulations using big data characteristics. Soft. Comput.Comput. **25**, 7527–7557 (2021)
60. Kumari, M.: A review on metaheuristic algorithms: recent and future trends. In: Metaheuristics-Based Materials Optimization, pp. 103–128 (2025)
61. Nabi, H.A., Faraj, K.H.A.: Enhanced classification of web services using hybrid meta-heuristic algorithms and deep learning. Expert Syst. Appl. **279**, 127281 (2025)
62. Gao, H., Kou, G., Liang, H., Zhang, H., Chao, X., Li, C.-C., Dong, Y.: Machine learning in business and finance: a literature review and research opportunities. Financ. Innov. **10**(1), 86 (2024)
63. Tam, J.H., Ong, Z.C., Ismail, Z., Ang, B.C., Khoo, S.Y.: A new hybrid GA–ACO–PSO algorithm for solving various engineering design problems. Int. J. Comput. Math.Comput. Math. **96**(5), 883–919 (2019)

Chapter 12
Application of Multi-criteria Decision-Making Methods

Abstract The chapter presents various methods of MCDM for use in decision-making situations with many factors. Each technique has its own way of structuring preferences and making decisions. We explain four main methods in this field, and use examples to demonstrate their practical use. Examples can be based on actual case studies, simulation work, or real data sets to make their use clear. Using more than one method for the same problem allows us to check the results and use what we learn to aid our decisions.

Learning Objectives

1. Learn how to use Fuzzy Delphi to identify important criteria based on expert opinion while taking ambiguity into account in evaluation.
2. Introduced a practical example explaining the steps involved in using the AHP to weight criteria.
3. Increase the skill of applying the EDAS method to rank alternatives based on analyzing distances from reference mean values.
4. Used MULTIMOORA method to rank alternatives by integrating three evaluation ways, providing accuracy and stability in decisions.

12.1 Selection Criteria Method

12.1.1 Fuzzy Delphi Analysis

Fuzzy Delphi Analysis (FDM) is a group decision-making tool that integrates the Delphi method, which harnesses the opinions of experts in a sequential manner with fuzzy logic to handle imprecision in decision-making opinions. FDM enables the feedback of the experts in linguistic terms (e.g., likely or important) that are then quantified in numerical terms through the use of fuzzy numbers, and thus facilitate a more accurate assessment where information is imprecise or subjective. This method

X. Chew et al., *Partial Least Squares Structural Equation Modeling and Complementary Methods in Business Research*, Information Systems Engineering and Management 67, https://doi.org/10.1007/978-3-032-01055-1_12

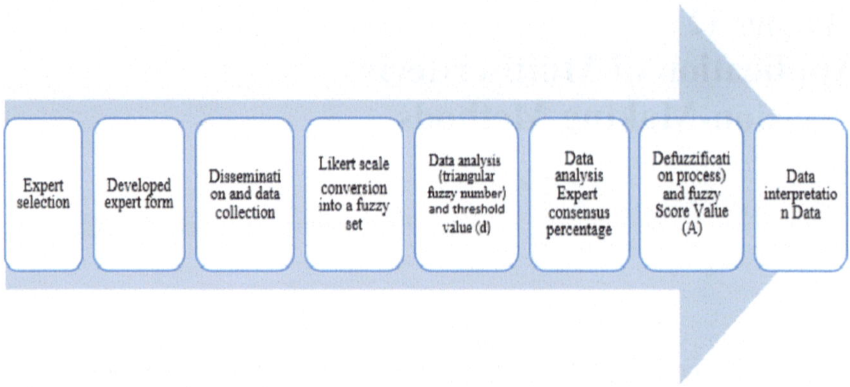

Fig. 12.1 Flowchart of the FDM approach

is particularly helpful in cases where group decision-making is needed, and the information available is vague. Figure 12.1 lays out the eight stages that makeup FDM.

Stage 1: Expert choosing is a major activity. Each expert panelists' judgment must be valid and appropriate to the research at hand. Therefore, picking the right expert is vital. Since FDM relies on an agreement between experts, a further challenge is settling on a sufficient number of experts. Previous research has shown that a sample size of 10–15 experts is optimal, particularly when there is substantial agreement among experts. At the same time, It's recommended that the panel consist of 10–50 experts [1–3].

Stage 2: Creating the specialist survey:

There are many options for creating the expert questionnaire, including (1) reviewing relevant literature and (2) conducting interviews. High flexibility makes it simpler to collect data from specialists and includes them in the research process, as claimed [4, 5].

Stage 3: Data collecting and distribution

The following methods may be used to disseminate and collect information from professionals: Two methods were used to distribute the questionnaire: (1) face-to-face meetings with each expert and (2) emailing and publishing the URL of the questionnaire on official network sites [6].

Stage 4: In this stage, the linguistic variables are transformed into triangular fuzzy numbers based on the aggregated findings from all experts. The linguistic variables for five Likert scales are shown in Table 12.1, which have been used in many studies [2, 7].

Stage 5: Based on the triangular fuzzy number, data analysis in this step aims to obtain the threshold value (d). The threshold value (d) must be less than or equal to

Table 12.1 Likert scale linguistic

Likert scale	Linguistic terms	Fuzzy numbers
1	Strongly disagree	(0.0, 0.0, 0.2)
2	Disagree	(0.0, 0.2, 0.4)
3	Less agree	(0.2, 0.4, 0.6)
4	Agreed	(0.4, 0.6, 0.8)
5	Strongly agree	(0.6, 0.8, 1.0)

0.2 ($d \leq 0.2$) as the first requirement of acceptance [8, 9]. The distance between the two fuzzy numbers is $m = (m_1, m_2, m_3)$, and $n = (n_1, n_2, n_3)$ can be computed by utilising the vertex method, whose formula is shown in the Eq. 12.1.

$$d(\tilde{m}, \tilde{n}) = \sqrt{\frac{1}{3}\left[(m_1 - n_1)^2 + (m_2 - n_2)^2 + (m_3 - n_3)^2\right]} \tag{12.1}$$

Stage 6: Expert consensus is calculated in this step; consensus or stability is generally agreed to be the fundamental aim of Delphi research. Consensus in a Survey instrument may be reached after two or more iterations of questionnaires. In light of this, it is crucial at this stage to get the consent of the panel of experts. The threshold for approval is set at 75% [7, 10].

Stage 7: In order to get the fuzzy score A (see Eq. 12.2), step 7 employs defuzzification. In order for A to satisfy the third criterion for admission, its value must be larger than or equal to the median alpha level (—cut value) of 0.5 [11]. Furthermore, A could be utilised to determine the element priority based on the views of others. The following equation can be employed to implement the fuzzy score A.

$$A = (1/3) \times (m_1 + m_2 + m_3) \tag{12.2}$$

The score A is employed to discover the crisp output of every fuzzy number. The centroid technique is employed for defuzzification as outlined in Eq. (12.2).

Stage 8: Analyse the results of the Fuzzy Delphi.

Here, the data is analysed using the fuzzy Delphi method of accepting or rejecting answers. The following are the three key requirements that must be met in order for an item to be erased or retained and for expert consensus to be reached [12, 13].

First requirement: Assess the experts' agreement on the performed element separately based on the fuzzy triangular numbers. A threshold value ($\leqq 0.2$) is required for acceptance [8].

Second requirement: If the majority level for consensus decision-making is more than or equal to 75%, then the conducted item will be considered [10].

Third requirement: A pseudo division, constructed with the aid of an expert's advice, may be used to determine the relative weight given to certain aspects of an object. The incorrect division may be prevented by establishing a threshold.

The cutoff value used here was a α-value of 0.5. Since 0.5 is in the middle of the interval, it was chosen as the value to utilise (0, 1). Therefore, this result backs up the reasonable argument that only items with sufficiently big' group grades are included in the fuzzy support set [14, 15]. Keep in mind that the summation of weight must be 1.

12.2 MCDM Weight Methods

12.2.1 AHP Method Application

The Analytic Hierarchy Process (AHP) is widely applied in decision-making scenarios such as project selection, resource allocation, and supplier evaluation by breaking down complex problems into simpler, hierarchical structures [16]. It facilitates group decision-making by enabling stakeholders to express their preferences and reach a consensus through pairwise comparisons of criteria and alternatives [17]. The first step of AHP is to prepare the questionnaire.

Questionnaire Questions Comparison Questions: Please put a circle on the (9 scales) to compare the services, as shown in the Table 12.2 below.

Step 1: Fill in the values obtained from the questionnaire in Table 12.2. Below is an example of how to get the decision matrix (see Table 12.3).
Step 2: Perform Normalization to make the data more similar in case of outliers.

Table 12.2 Nine scales of pairwise comparisons

Intensity of importance	Definition
1	Equal importance
3	Weak importance of one over another
5	Essential or strong importance
7	Demonstrated importance
9	Absolute importance
2, 4, 6, 8	Intermediate values between the two adjacent judgments

Table 12.3 Decision matrix for the AHP method

Performance		Original DM			
		Trust	Innovation	Leadership	Integrity
First expert	Trust	1.000	2.000	3.000	1.000
	Innovation	0.500	1.000	3.000	4.000
	Leadership	0.333	0.333	1.000	2.000
	Integrity	1.000	0.250	0.500	1.000

For example, data was collected on a scale from 1 to 9, but one of the experts mistakenly gave a number of 90 for one of the criteria. Below is an example of how to get the normalization.

Calculate the Column Sums: Referring to your pairwise comparison matrix you must add up all values inside each individual column. The step produces the total value for each criterion by examining values across every comparison between the criteria [18]. For example, the summation of the first column is $(1 + 0.5 + 0.333 + 1) = 2.83$.

Normalize Each Element: Each entry in the matrix should be divided by the total in its respective column. Each entry receives transformation into a value ratio that represents its total from its column amount. The normalized matrix demonstrates a sum of 1 for each column. For example, the first value of the first criterion is 0.353 were comes from this equation $1/2.83 = 0.353$. Where 1 comes from the first criterion in the decision matrix, and 2.83 is the result of the summation for the first criterion. Table 12.4 Shows the normalization decision matrix.

Step 3: The normalized values are calculated based on the SUM equation. The second normalization is the horizontal summation of each alternative as example below.

$\text{Aggr} = (0.353 + 0.558 + 0.400 + 0.125) = 1.436$ as shown in Table 12.5

Step 4: The weight is calculated, which is the result of dividing Aggregation by the sum of the criteria as shown in Table 12.5.

Keep in mind that the summation of weight must be 1.

Table 12.4 Normalization decision matrix

Normalized DM			
Trust	Innovation	Leadership	Integrity
0.353	0.558	0.400	0.125
0.176	0.279	0.400	0.500
0.118	0.093	0.133	0.250
0.353	0.070	0.067	0.125

Table 12.5 Final weighting

Normalized DM				Aggregation	Final weight
Trust	Innovation	Leadership	Integrity		
0.353	0.558	0.400	0.125	1.436	1.436/4 = 0.359
0.176	0.279	0.400	0.500	1.356	0.339
0.118	0.093	0.133	0.250	0.594	0.148
0.353	0.070	0.067	0.125	0.614	0.154
SUM					1

12.3 MCDM Ranking Methods

12.3.1 EDAS Method Application

One of the more effective MCDM ranking method named, Evaluation Based on Distance from Average Solution (EDAS) measures alternative attractiveness by identifying the distance between alternatives and an average criterion solution [19]. EDAS operates differently than standard methods by using average calculations to measure distances from average series which determine alternative desirability based on positive and negative values [20]. Research demonstrates that EDAS successfully helps manage large inventory control systems when multiple criteria exist simultaneously [21]. Furthermore, the method supports hiring decisions for construction managers to effectively select personnel in the building sector. Due to its strong resistance and flexible nature EDAS has been integrated into numerous disciplines such as agriculture, business management, and healthcare which creates it an essential instrument for decision-makers who need to handle problems with conflicting criteria. The steps of EDAS is shown below [22]. Let's assume we have four criteria and five alternatives, and we also weight each criterion as shown in the Table 12.6.

Step 1: Determine the average solution (AV_j)

$$AV_j = \frac{\sum_{i=1}^{n} X_{ij}}{n} \tag{12.3}$$

where n is the number of alternatives.

Step 2: Calculate the positive distance from average (PDA) if j th criterion is beneficial,

$$PDA_{ij} = \frac{\max(0, (X_{ij} - AV_j))}{AV_j} \tag{12.4}$$

Table 12.6 The example of how to find (AV_j)

Weightage	Non-Beneficial	Beneficial	Beneficial	Beneficial
	0.350	0.250	0.250	0.150
	Price/Cost	Storage space	Camera	Looks
M1	250	16	12	5
M2	200	16	8	3
M3	300	32	16	4
M4	275	32	8	4
M5	225	16	16	2
AV_j	= sum (250,200,300,275,224)/5 = **250**	**22.4**	**12**	**3.6**

if j th criterion is non-beneficial,

$$PDA_{ij} = \frac{\max\left(0, \left(AV_j - X_{ij}\right)\right)}{AV_j} \tag{12.5}$$

The result of Eqs. (12.4, 12.5) is shown in Table 12.7 below.

Step 3: Calculate the negative distance from average (NDA)

if j th criterion is beneficial,

$$NDA_{ij} = \frac{\max\left(0, \left(AV_j - X_{ij}\right)\right)}{AV_j} \tag{12.6}$$

if j th criterion is non-beneficial,

$$NDA_{ij} = \frac{\max\left(0, \left(X_{ij} - AV_j\right)\right)\Big)}{AV_j} \tag{12.7}$$

The result of Eqs. (12.6, 12.7) is shown in Table 12.8 below.

Step 4: Weighted sum of PDA

Table 12.7 Positive distance from average (PDA)

Weightage	0.35	0.25	0.25	0.15
	Price/Cost	Storage space	Camera	Looks
M1	0.000	MAX (0;(16 – 22.4))/22.4 = 0	0.000	0.388
M2	MAX (0;(250 – 200))/250 = 0.2	0.000	0.000	0.000
M3	0.000	0.428	0.333	0.111
M4	0.000	0.428	0.000	0.111
M5	0.100	0.000	0.333	0.000

Table 12.8 Negative distance from average (NDA)

Weightage	0.350	0.250	0.250	0.150
	Price/Cost	Storage space	Camera	Looks
M1	– MAX(0;(250 – 250))/250 = 0	= MAX(0;(22.4 – 16))/22.4 = 0.285	0.000	0.000
M2	0.000	0.286	0.333	0.166
M3	0.200	0.000	0.000	0.000
M4	0.100	0.000	0.333	0.000
M5	0.000	0.285	0.000	0.444

$$SP_i = \sum_{j=1}^{m} w_j PDA_{ij} \tag{12.8}$$

Depending on Table 12.6 and Eq. 12.8, the SP_i will product as shown in Table 12.9.

Step 5: Weighted sum of NDA

$$SN_i = \sum_{j=1}^{m} w_j NDA_{ij} \tag{12.9}$$

Depending on Table 12.7 and Eq. 12.9. the SN_i will product as shown in Table 12.10.

Step 6: Calculate Normalize the values of SP_i and SN_i

$$NSP_i = \frac{SP_i}{\max_i(SP_i)}$$
$$NSN_i = 1 - \frac{SN_i}{\max_i(SN_i)} \tag{12.10}$$

Step 7: Normalize the values of NSP and NSN

$$AS_i = \frac{1}{2}(NSP_j + NSN_i) \tag{12.11}$$

Table 12.9 SP_i

Price/Cost	Storage space	Camera	Looks	SP_i
0.350 * 0 = 0	0.000	0.000	0.058	SUM(0, 0, 0, 0.0583) = 0.058
0.350 * 0.2 = 0.070	0.000	0.000	0.000	SUM(0.07, 0, 0, 0) = 0.070
0.35 * 0 = 0	0.107	0.083	0.017	0.207
0.35 * 0 = 0	0.107	0.000	0.017	0.123
0.350 * 0.1 = 0.035	0.000	0.083	0.000	0.118

Table 12.10 SN_i

Price/Cost	Storage space	Camera	Looks	SN_i
0.350 * 0 = 0	0.071	0	0	SUM (0, 0.071, 0, 0) = 0.071
0.350 * 0 = 0	0.071	0.083	0.025	SUM (0, 0.071, 0.083, 0.025) = 0.179
0.350 * 0.2 = 0.070	0	0	0	0.070
0.350 * 0.100 = 0.035	0	0.083333	0	0.118
0.350 * 0 = 0	0.071	0	0.066	0.138

Table 12.11 Final rank for EDAS method

Options	SP_i	SN_i	NSP_i	NSN_i	AS_i	Rank
M1	*0.058*	0.071	*0.058/0.207 = 0.281*	*0.602*	0.5 * (0.281 + 0.602 = 0.442	**3**
M2	0.070	0.180	0.338	0	0.169	**5**
M3	0.207	*0.070*	1	1 − (0.07/ 0.179) = 0.610	0.805	**1**
M4	0.124	0.118	0.598	0.342	0.470	**2**
M5	0.118	0.138	0.571	0.232	0.402	**4**
max	0.207	0.179				

The final rank of alternatives is shown in Table 12.11.

12.3.2 Multimoora

MULTIMOORA stands as a powerful MCDM technique combining the ratio system with full multiplicative form along with the reference point approach under the acronym "Multi-Objective Optimization on the basis of Ratio Analysis plus the Full Multiplicative Form." The MULTIMOORA system merges MOORA with three evaluation procedures including the ratio system and the reference point approach and the full multiplicative form. The individual components in this system produce unique performance assessments which enable better insights beyond what single method evaluation provides [23, 24].

The ratio procedure collects weighted measurements of performances across alternatives by combining beneficial weightings with beneficial criteria while simultaneously combining non-beneficial criteria weightings. The initial performance index obtained from this method evaluates alternative effectiveness toward the accomplishment of desired objectives. Through the Min–Max distance-based metric the reference point approach measures deviations of alternatives from a perfect solution. The procedure both reduces personal interpretation and demonstrates that effective solutions must maintain proximity to highest possible results on every measured criterion [25].

These methods find completion in the full multiplicative form because it tackles potential interdependencies that exist between different criteria. When performance ratios are multiplied throughout this process the procedure provides improved analytical detail which accounts for multi-dimensional performance measurement interactions. Such conceptualization proves most helpful when changes in single criteria result in noticeable modification of final assessment since it maintains crucial elements in focus. When MULTIMOORA unites dominating elements from three methodologies it forms a comprehensive ranking which presents an equilibrium between compensatory and non-compensatory decision elements. The MULTIMOORA system provides decision-makers with a capability to manage both quantitative and qualitative criteria evaluation processes which ensures suitable evaluation of alternatives with diverse measurement units. The ability of MULTIMOORA to combine multiple evaluation perspectives creates a strong and dependable system which aids decision-making in challenging scenarios where different goals compete against each other while information remains uncertain [26].

Let our matrix is contain 5 criteria and 4 alternatives as shown in Table 12.12.

The first step is normalization the value by using Eq. 12.12.

$$NX = \frac{x}{\sqrt{\sum_{j=1}^{m} x^2}} \tag{12.12}$$

The generation process of values appears in example below which defines and represents the numerous elements that construct the value formation. The next example demonstrates the fundamental components and relationships between elements as well as the elements that lead to the final computed value by using mathematical models.

$$\begin{aligned} \text{Sum of Squares of each column} &= (200)^2 + (250)^2 + (150)^2 \\ &\quad + (300)^2 = 21500000000000 \\ \text{SQRT of } (21500000000000) &= 4636809 \end{aligned}$$

$NX = 200/4{,}636{,}809 = 0.432$ as shown in first column in Table 12.13. The final result of the normalization is shown in Table 12.13.

The Ratio System evaluates alternative performances on different criteria points by calculating comprehensive scores. A composite score is calculated through the process that combines beneficial criterion values meticulously normalized to higher amounts and subtracts non-beneficial criterion values normalized to lower amounts (see Eq. 12.13). Higher scores produced from the composite score help decision-makers make objective rankings because they determine more favorable options. Tables 12.14 and 12.15 shown the result of Ratio System.

Table 12.12 Decision matrix for the MULTIMOORA method

Options	C1	C2	C3	C4	C5
A1	200	4	2	2	5
A2	250	5	2	2	9
A3	150	4	5	1	7
A4	300	8	3	1	9

Table 12.13 Normalization

Options	Cost	Benefit	Benefit	Benefit	Benefit
	C1	C2	C3	C4	C5
A1	0.432	0.364	*2/6.480 = 0.308*	0.632	*5/15.362 = 0.325*
A2	0.539	0.455	0.309	0.632	*9/15.362 = 0.585*
A3	0.323	0.364	0.772	0.316	0.456
A4	0.647	0.727	0.463	0.316	0.586

$$y_i = \sum_{B_j \in B_{\text{benffit}}} \varpi_j r_{ij} - \sum_{B_j \in B_{\text{coas}}} \varpi_j r_{ij} \quad i = 1, 2, \ldots, m \tag{12.13}$$

The Ratio System in MULTIMOORA generates results which display the performance evaluation values of four alternatives after normalization and weighting of criteria. Performance scores in the Ratio System measure collective beneficial and non-beneficial factor relationships and translate into superior results. The first rank goes to the alternative with highest score (0.371) and the other alternatives receive scores ranking them second (0.289), third (0.269) and last (0.203). The decision-maker's objectives match the first option according to this ranking method with the fourth option delivering the most unfavorable results among the considered options.

The Reference Point approach of the MULTIMOORA method decides alternative rankings through the measurement of their distance to an ideal solution designated as reference point (see Eqs. 12.15 and 12.16). The reference point uses ideal values from selected best performances that normalize criteria into maximum benefactors

Table 12.14 How to calculate ratio system

Weight	0.200	0.100	0.300	0.100	0.300
	Cost	Benefit	Benefit	Benefit	Benefit
	C1	C2	C3	C4	C5
A1	0.2 * 0.431 = 0.086	0.1 * 0.363 = 0.036	0.093	0.063	0.098
A2	0.2 * 0.539 = 0.107	0.045	0.093	0.063	0.176
A3	0.065	0.036	0.231	0.032	0.137
A4	0.129	0.073	0.139	0.032	0.176

Table 12.15 Ratio system

Ny	Ratio system
(0.036 + 0.092 + 0.063 + 0.097) − 0.086 = *0.203*	4
0.269	3
0.371	1
0.289	2

and minimum non-benefactors. The reference point values are established during normalization then alternatives receive an absolute deviation evaluation. The ranking order follows a system that determines maximum deviations from reference points which identifies the best alternatives as those demonstrating minimal distance from reference benchmarks. The technique establishes nearness to an optimal benchmark for evaluating alternatives in a clear and unbiased way. The first step of finding Reference point is shown in Table 12.16.

$$t_j = \begin{cases} \max\limits_{B_j \in B_{\text{cax}}} r_{ij} \\ \min\limits_{B_j \in B_{b\,\text{cenfi}}} r_{ij} \end{cases} \tag{12.14}$$

$$z_i = \frac{1}{n} \sum_{i=1}^{n} d_{ij} \quad i = 1, 2, \ldots, m \tag{12.15}$$

After multiplying the original values of the matrix by the weights for each criterion, we must find the smallest and largest values depending on whether the criterion was a benefit or a cost (see Table 12.17).

The final step of reference point is to find the max value of each alternative as shown in Table 12.18.

Table 12.18 Final step to find Reference Point.

Reference Point results determine the maximum distance between each alternative solution and its optimal solution point and lower numbers indicate superior match with optimum criteria standards. The alternative with 0.039 stands as first in ranking based on its minimal distance from the perfect reference standard. Alternative number 2 achieves a ranking of second place with a deviation value of 0.092. Among all

Table 12.16 First step to find reference point

Weight	0.200	0.100	0.300	0.100	0.300
	Cost	Benefit	Benefit	Benefit	Benefit
	C1	C2	C3	C4	C5
A1	0.2 * 0.431 = 0.086	0.1 * 0.366 = 0.036	0.093	0.063	0.098
A2	0.2 * 0.539 = 0.107	0.045	0.093	0.063	0.176
A3	0.064	0.036	0.231	0.032	0.137
A4	0.129	0.073	0.139	0.032	0.176

Table 12.17 Second step to find reference point

Reference point	= MIN (*0.086*, 0.107, 0.064, 0.129) = *0.064*	= MAX (0.036, 0.045, 0.036, 0.073) = 0.072	= MAX (0.093, 0.093, 0.231, 0.139) = 0.231	= MAX (0.063, 0.063, 0.032, 0.032) = 0.063	= MAX (0.098, 0.176, 0.137, 0.176) = 0.1757

Table 12.18 First step to find full multiplicative

Weight	0.200	0.100	0.300	0.100	0.300
	Cost	Benefit	Benefit	Benefit	Benefit
	C1	C2	C3	C4	C5
A1	0.2 * 0.4313311 = 0.086	0.1 * 0.363 = *0.036*	*0.092*	*0.063*	*0.097*
A2	0.2 * 0.539 = 0.107	0.045	0.092	0.063	0.175
A3	0.064	0.036	0.231	0.031	0.136
A4	0.1293	0.072	0.138	0.031	0.175

alternatives two achieve the maximum deviation of 0.138 at the same time and thus obtain equal 3rd rank positions.

The Full Multiplicative Form functions as the third element of the MULTIMOORA approach which operates together with Ratio System and Reference Point for multi-criteria decision assessments (see Eq. 12.16). The method determines alternative scores through multiplying normalized beneficial criterion assessments and then splitting the outcome by non-beneficial criterion assessments. Tables 12.18 and 12.19 show the result.

$$u_i = \frac{\prod_{B_j \in B_{\text{bexfit}}} (r_{ij})^{\sigma_j}}{\prod_{B_j \in B_{\text{cost}}} (r_{ij})^{\sigma_j}} \tag{12.16}$$

After multiplying the original values of the matrix by the weights for each criterion, Column A is the product of multiplying all the benefit criteria, column B is the product of the cost criteria, and column U is the division of A/B. Table 12.19 shows the result of full multiplicative.

The Full Multiplicative Form method calculates alternative performance by dividing beneficial criteria products by non-beneficial criteria products to obtain utility values (U). The alternative achieving the maximum U value at 0.0005holds the first rank as the top-performing choice. The alternative with the U value of 0.0004 ranks along with another option as 2nd and 3rd followed by the 4th place option which has a U value of 0.0002. The utility values allow for the most efficient ranking of alternatives since higher utility indicates a superior ratio of benefits to costs in multiplicative assessments.

Table 12.19 Final step to find full multiplicative

A	B	U	Multiplicative form
(0.036 * 0.092 * 0.063 * 0.097) = 2.07902E−05	0.086	(2.07902E−05/0.086) = 0.0002	4
4.6778E−05	0.107	0.0004	2
3.63829E−05	0.064	0.0005	1
5.61336E−05	0.129	0.0004	3

Table 12.20 Final rank for MULTIMOORA method

Options	Ratio system	Reference point	Multiplicative form	Final rank
A1	4	3	4	4
A2	3	3	2	3
A3	1	1	1	1
A4	2	2	3	2

The final stage of MULTIMOORA method is collected all result in on table to get the final rank. Table 12.20 is shown the final rank.

The final ranking table presents performance of four alternatives (A1–A4) based on all three components of MULTIMOORA methodology: Ratio System and Reference Point together with Full Multiplicative Form. The final rank of alternatives A1–A4 emerges from aggregating their total rankings obtained in each of the three methods through majority or consensus procedures. The MULTIMOORA evaluation system shows A3 standing at the top position in every scoring method which confirms its selection as the best alternative. A4 achieves second position through the Ratio System and Reference Point followed by third in multiplicative form to secure second place in the FINAL RANK. A2 obtains a final rank of 3rd although it places 3rd in both methods and scores a 2nd position in the third method. Among the alternatives A1 holds the lowest rank because it achieves position 4 in Ratio System and Multiplicative Form and position 3 in reference point evaluation thus obtaining final position 4. When different methods merge into one evaluation system it leads to an improved and balanced method of decision-making.

References

1. Adler, M., Ziglio, E.: Gazing into the Oracle: The Delphi Method and Its Application to Social Policy and Public Health. Jessica Kingsley Publishers (1996)
2. Manakandan, S.K., Rosnah, I., Mohd, R.J., Priya, R.: Pesticide applicators questionnaire content validation: a fuzzy delphi method. Med. J. Malaysia **72**(4), 228–235 (2017)
3. Parsazadeh, N., Ali, R., Rezaei, M., Tehrani, S.Z.: The construction and validation of a usability evaluation survey for mobile learning environments. Stud. Educ. Eval. **58**, 97–111 (2018)
4. Moradpour, N., Pourahmad, A., Ziari, K., Hataminejad, H., Sharifi, A.: Downscaling urban resilience assessment: a spatiotemporal analysis of urban blocks using the fuzzy Delphi method and K-means clustering. Build. Environ. **263**, 111898 (2024)
5. Powell, C.: The Delphi technique: myths and realities. J. Adv. Nurs. **41**(4), 376–382 (2003)
6. Abdullah, J.B.B., Yusof, S.: A fuzzy Delphi method-developing high-performance leadership standard for Malaysian school leaders. J. Educ. Soc. Sci. **9**(2), 1–10 (2018)
7. Dawood, K.A., Sharif, K.Y., Ghani, A.A., Zulzalil, H., Zaidan, A.A., Zaidan, B.B.: Towards a unified criteria model for usability evaluation in the context of open source software based on a fuzzy Delphi method. In: Information and Software Technology, vol. 130. (2021). https://doi.org/10.1016/j.infsof.2020.106453
8. Cheng, C.-H., Lin, Y.: Evaluating the best main battle tank using fuzzy decision theory with linguistic criteria evaluation. Eur. J. Oper. Res. **142**(1), 174–186 (2002)

9. Roldan Lopez de Hierro, A.F., Sánchez, M., Puente-Fernández, D., Montoya-Juárez, R., Roldán, C.: A fuzzy delphi consensus methodology based on a fuzzy ranking. Mathematics **9**(18), 2323 (2021)
10. Chu, H.-C., Hwang, G.-J.: A Delphi-based approach to developing expert systems with the cooperation of multiple experts. Expert Syst. Appl. **34**(4), 2826–2840 (2008)
11. Tang, C.-W., Wu, C.-T.: Obtaining a picture of undergraduate education quality: a voice from inside the university. High. Educ. **60**(3), 269–286 (2010)
12. Abdulkareem, K.H., Arbaiy, N., Zaidan, A.A., Zaidan, B.B., Albahri, O.S., Alsalem, M.A., Salih, M.M.: A new standardisation and selection framework for real-time image dehazing algorithms from multi-foggy scenes based on fuzzy Delphi and hybrid multi-criteria decision analysis methods. Neural Comput. Appl. **33**(4), 1029–1054 (2021)
13. Dawood, K.A., Sharif, K.Y., Ghani, A.A., Zulzalil, H., Zaidan, A.A., Zaidan, B.B.: Towards a unified criteria model for usability evaluation in the context of open source software based on a fuzzy Delphi method. Inf. Softw. Technol. **130**, 106453 (2021)
14. Al-Rikabi, Y.K., Montazer, G.A.: Designing an e-learning readiness assessment model for Iraqi universities employing Fuzzy Delphi Method. Educ. Inf. Technol. **29**(2), 2217–2257 (2024)
15. Bodjanova, S.: Median alpha-levels of a fuzzy number. Fuzzy Sets Syst. **157**(7), 879–891 (2006)
16. Darko, A., Chan, A.P.C., Ameyaw, E.E., Owusu, E.K., Pärn, E., Edwards, D.J.: Review of application of analytic hierarchy process (AHP) in construction. Int. J. Constr. Manag. **19**(5), 436–452 (2019)
17. Canco, I., Kruja, D., Iancu, T.: AHP, a reliable method for quality decision making: a case study in business. Sustainability **13**(24), 13932 (2021)
18. Buran, B., Erçek, M.: Public transportation business model evaluation with Spherical and Intuitionistic Fuzzy AHP and sensitivity analysis. Expert Syst. Appl. **204**, 117519 (2022)
19. Hashemkhani Zolfani, S., Ebadi Torkayesh, A., Ecer, F., Turskis, Z., Šaparauskas, J.: International market selection: a MABA based EDAS analysis framework. Oeconomia Copernicana **12**(1), 99–124 (2021)
20. Torkayesh, A.E., Deveci, M., Karagoz, S., Antucheviciene, J.: A state-of-the-art survey of evaluation based on distance from average solution (EDAS): developments and applications. Expert Syst. Appl. **221**, 119724 (2023)
21. Ren, J., Hu, C., Yu, S., Cheng, P.: An extended EDAS method under four-branch fuzzy environments and its application in credit evaluation for micro and small entrepreneurs. Soft. Comput. **25**(4), 2777–2792 (2021)
22. Paul, V.K., Chakraborty, S., Chakraborty, S.: An integrated IRN-BWM-EDAS method for supplier selection in a textile industry. Decis. Making: Appl. Manage. Eng. **5**(2), 219–240 (2022)
23. Ghaemi-Zadeh, N., Eghbali-Zarch, M.: Evaluation of business strategies based on the financial performance of the corporation and investors' behavior using D-CRITIC and fuzzy MULTI-MOORA techniques: a real case study. Expert Syst. Appl. **247**, 123183 (2024)
24. Zhang, X., Li, J., Wang, N.: Sustainable location selection for photovoltaic waste recycling centers: an extended MULTIMOORA approach. Sustain. Cities Soc. 106316 (2025)
25. Lin, M., Huang, C., Xu, Z.: MULTIMOORA based MCDM model for site selection of car sharing station under picture fuzzy environment. Sustain. Cities Soc. **53**, 101873 (2020)
26. Göçer, F., Büyüközkan, G.: A novel extension of Pythagorean fuzzy MULTIMOORA approach for new product development. Heliyon **9**(6) (2023)

The manufacturer's authorised representative in the EU is Springer Nature Customer Service Centre GmbH, Europaplatz 3, 69115 Heidelberg, Germany. If you have any concerns regarding our products, please contact ProductSafety@springernature.com

Printed and bound by CPI Group (UK) Ltd, Croydon, CR0 4YY
01/10/2026
02225864-0002